IMAGINARY DEATH

Before you start to read this book, take this moment to think about making a donation to punctum books, an independent non-profit press,

@ https://punctumbooks.com/support/

If you're reading the e-book, you can click on the image below to go directly to our donations site. Any amount, no matter the size, is appreciated and will help us to keep our ship of fools afloat. Contributions from dedicated readers will also help us to keep our commons open and to cultivate new work that can't find a welcoming port elsewhere. Our adventure is not possible without your support.

Vive la Open Access.

Fig. 1. Detail from Hieronymus Bosch, *Ship of Fools* (1490–1500)

IMAGINARY DEATH. Copyright © 2025 by Mariko Nagai. This work carries a Creative Commons BY-NC-SA 4.0 International license, which means that you are free to copy and redistribute the material in any medium or format, and you may also remix, transform, and build upon the material, as long as you clearly attribute the work to the author (but not in a way that suggests the author or punctum books endorses you and your work), you do not use this work for commercial gain in any form whatsoever, and that for any remixing and transformation, you distribute your rebuild under the same license. http://creativecommons.org/licenses/by-nc-sa/4.0/

Published in 2025 by punctum books, Earth, Milky Way.
https://punctumbooks.com

ISBN-13: 978-1-68571-236-5 (print)
ISBN-13: 978-1-68571-237-2 (ePDF)

DOI: 10.53288/0531.1.00

LCCN: 2025942579
Library of Congress Cataloging Data is available from the Library of Congress

Editing: Vincent W.J. van Gerven Oei and SAJ
Book design: Hatim Eujayl
Cover image: Shirō as a young man, 1933.
Cover design: Vincent W.J. van Gerven Oei

spontaneous acts of scholarly combustion

HIC SVNT MONSTRA

Mariko Nagai

Imaginary Death

p.

Contents

Prelude · 17

I. The Vague Uneasiness · 21
II. The Chosen One · 65
III. The Sacred War · 103
IV. Domestic Life · 155
V. The War without an End · 197

Epilogue · 257

Bibliography · 277

Acknowledgments

I am grateful for the following art foundations and residencies for their belief in this work as well as for providing time and space (and amazing food): Akademie Schloss Solitude; le Chateau de Lavigny Writers' Residency and the Ledig-Rowohlt Foundation; Djerassi Resident Artists Program; Fundacion Valparaiso; the International Retreat for Writers at Hawthornden Castle; PEN-Schrijversflat and PEN Vlaanderen; the Rockefeller Foundation Bellagio Center; and Yaddo. Temple University Japan Campus, my home for the past twenty-five years, has continued to support me in tangible and intangible ways.

Excerpts of this work appeared in *Asia Literary Review, Third Coast, New Letters, Southern Humanities Review,* and *Memoire(and).*

I have conducted online and/or onsite research in the following archives and libraries: National Archives and Records Administration (USA), the National Institute for Defense Studies Military Archives (Japan), Yasukuni Shrine Archive (Japan), Chiran Tokko Museum (Japan), Seijo University Library (Japan), Showa Women's College Library (Japan), Saratoga Springs Public Library (USA), and the New York Public Library (USA).

The following individuals have extended their hearts and friendship and offered encouragement and help while I struggled and nearly drowned so many times in the gravity and futility of this work: my mother's aunts and uncle — Aunts Aiko and Asako, who found my grandfather's letters in a dark corner of a shed and sent them to me after my inquiry and who shared their memories that would have otherwise been lost if it were not for both of them, and Uncle Tsutomu (d. 2018), the only surviving brother, who presented me with my grandfather's journal in China — I know you gave me the most precious thing in your life, and I am grateful for that; Jonathan Wu (d. 2022), whom I talked to on an everyday basis, obsessively, insistently, about small findings over late-night dinners after work, and who only offered encouragement and kind ears and suggestions — a best kind of best friend anyone can have; the fellows at the Rockefeller Foundation Bellagio Center, who, in the truest sense, created a sense of fellowship during our month together; Silvia Pareschi, who always makes me laugh but who understands, more than anyone, what it means to be a writer; Amy Kirsten, who offered that lovely lunch in Manhattan on a cold February afternoon; Ingrid Vander Veken and Paul, her partner, who opened their home for me in Antwerp as well as translated my work into Dutch; and my mother (d. 2020), who was instrumental in getting letters and the journal from my relatives, and who tirelessly told me stories of her family over dinner, breakfast, and tea. After so many losses, I know, keenly, what Yevtushenko means: "worlds die in them."

For My Mother (1941–2020),
who never stopped missing her father whom she never knew.

These are your stories.

*In any man who dies there dies with him
his first snow and kiss and fight.
[…]
Not people die but worlds die in them.*

—Yevgeny Yevtushenko, *"People"*

Prelude

July 8, 1944: A Man Dies, Where the Story Starts

He dies because he must — because without his death, there is no story, and, in the end, no history itself. And because history, by its nature, follows a linear time movement, it goes forward, leaving behind bodies of men and women whose worlds and dreams die when they die.

But here is a man. He has just turned twenty-eight half a month ago. He has three children and a wife back home in Japan. He is a sergeant with a dozen or so men under his command, and he is in New Guinea. He is a seasoned soldier: first, as the Emperor's Guard, then on a tour of duty in China for two years. His third and final tour of duty is on this green island where men die from hunger as quickly as seven-day cicadas, where they die from malaria and simple infections. Until here, he did not know that men die, most of all, from homesickness so keen that the soul disengages from its body even while not asleep to make its way back home. He has received a command, the first in eighteen months: to march toward the Driniumor river for the first and final offensive attack.

And in the next scene, he lies by the riverbank, dead. As he dies, he takes with him the memory of his first embrace, his love, his thoughts, his dreams, and all the hopes he never actualized during the life he lived — so short it could have been the mere sigh of a god.

Fig. 1.1. Susumu (left) and Shirō (right) circa 1930.

I

The Vague Uneasiness

June 26, 1916: The Birth of Shirō

Shirō is born amidst the war, during the reign of the weak Emperor Taishō. This century of his birth will know only two dozen days without war and battles. This century: of bodies tossed in piles, days measured by lives of men and women torn away from their hopes and unrealized dreams. World War I is not the first of the century. No. And it will not be the last one. Many will follow and push into the next millennium. But here is a baby, Shirō, *the man of seeds,* a fitting name for a farmer's son. An appropriate name for a man who would have been destined to work on land all his life. He is a third son on paper, though he is the first one to survive. His father, Saburō, *the third son,* married into his wife, Eiko's, family because he is a third son; dispensable, because he would not carry on the family land and line, because marrying into a woman's family and becoming the head of the family would allow him to avoid conscription, because at this time, in this place where tradition holds stronger than the modern age, a woman, if she is an only child, must have a man marry into the family. This is a joyous occasion: the first son. The surviving son. At this time, no one knows that Shirō's life will end with the war, as if his life is interwoven with the narrative of the war, as if his life is part of the history, unable to resist, unable to unchain himself from the world itself.

August 1916: The Farm, and Nothing Beyond

The hamlet is named Ōsu, or *Big Sandbank,* surrounded by two major rivers that it almost seems like a sandbank, though in reality, it's land-connected. Right near where Ibaraki and Chiba prefectures meet. So close to the sea that the area is called Itako, or the *Arrival of the Ocean Water.* A story goes: once upon a time, a long time ago when gods were still closer to men and men to gods, a god, Takekashima, came to rule this land by the sea. And how the people protested! They took arms, hoes, clubs, trying to drive away the god, but the god, angered by the insolence of these mere mortals, slaughtered each and every one of them, including the children and old men who carried handmade weapons in their hands, just like the men they were to become, or the men they once were. The phrase, *itaku koroshita,* or "killed to its heart's content," through time and mouths became Itako, and when men began to forget gods and written words came, they wrote *Arrival of the Ocean Water* as its name. That was many years ago, and for the last four hundred years, the town has been a major stop known for field after field of purple irises, for women captains punting down the river, for gay houses that never extinguished their red lanterns, where women leaned down from windows high above, waving hands to men on boats to lure them in, for white herons, for resilient farmers who revolted against the government again and again, and for the fierce swordsmanship of its warriors and violent resistance against injustice. A town where people kept boats more than horses, and men and women learned to swim even before they learned how to walk. On a damp day, with eyes closed, the scent of the sea can be smelled here. It is not a village, no. It is too small to be called a village. A hamlet composed of the same families, once removed, married into cousins, where someone from a hamlet over will always be an outsider. A river — one of many rivers running out of the major one like fingers of a hand — runs behind this farmhouse, a river that eventually connects with the sea downstream in a matter of one, two hours, and, upstream, so many days of navigating a boat, to the imperial capital. On a regular day, boats come and go, carrying cargo, passengers, vegetables, barrels of rice — a major artery during this time in history when roads are still rendered useless during rain and rivers are important not just for where they can take people but also for sustaining lives. Wooden boats, pushed by poles, come and go. You can spend the entire day watching boats carrying brides in black kimonos, modestly keeping their chins tilted down. They carry gods on festival days, they carry the bodies of the dead and the living and all their histories. The river rises. The river lowers itself. According to the weather. According to the time of the year or day. But the farm is consistent. Flat. The horizon, the shades of the hills, all flat, and you can see a house here, a house there, for miles and miles, until you can almost see the ocean to the east and the tall buildings in the capital to the southwest. But it is not idyllic; like any small hamlet in this archipelago, modernization has touched it, ever so slightly. The idyllic, the pastoral of the past, where men and women, unmarried and married, chose partners according to the diction of the area, is gone. The minor gods of the hamlet have been pushed aside by a greater god, named Emperor Taishō. Women can no longer choose to stuff their birthing canals with winter cherries to abort an unwanted fetus *because they cannot afford extra mouths*. It is now the time of modernity, a time where *progress* is synonymous with denial of the past. The nation is running toward modernization; men and women must be ready to answer the call of the nation, at any time, any place, and even in this small farmhouse of the landowner class, they are, and they will be. This is the land where, for men and women, water is both a blessing and a curse, and the land equally so.

Meiji, Then Taishō

To the West, the West, we look to the West. From shogun to emperor; from an island composed of clan lords under one shogun to a nation-state; from lives defined by one of the four classes of birth — and one that remained unnamed — to unchartered lives. As soon as the border opened up, men and women, after nearly 350 years of isolation, packed their bags and crossed oceans everywhere to explore, to study, to see, to re-forge their lives by their sheer will, or at least luck: manual laborers, circus troupes, students, immigrants, businessmen, diplomats on the strange shores of America, Russia, China, French Indochina, sometimes as far away as Africa and Europe. The nation, with an emperor instead of a shogun, sent out selected men with a command, *learn all you can from the West, and bring back information so we can be as good as the West.* And dutifully, men came back after many years of wandering and reported the wondrous sights they had seen, ideas they had studied, and the nation took note and copied what they heard. The nation demolished classes; it took away powers from clan lords and gave them aristocratic titles; they made men cut off their hair and changed the calendar, which changed the whole nature of season and time. Instead of samurai responsible for fighting, all men over twenty years old became eligible to fight, to serve, by saying, *A wealthy nation is made by strong soldiers.* The West, the West, we look to the West. But little did we know that those who have seen the West came back scarred, without knowing that they have been wounded. The nation takes all that they have seen, and applying all it has learned, it wages war against Russia and wins, *We are of the West, modern, we are equal to the West.* And now that the old era ended with the death of the Emperor, a new era, Taishō, named after a *Daodejing* saying, *After many ways, the only true way is the Way,* as if they have finally settled on who they are. Who are we, we ask ourselves. Who are we, are we of the West, better than the West, and if we are of the West, is it not our right to do what the West is doing — to expand, to colonize, to domesticate?

October 21, 1916: Birth of Masa

Several villages over, a girl is born, the third child in a wealthy landlord's family. Her name: Masa Ōkawa. She is a pale infant, so pale her skin seems to be made out of snow, a rare kind of flesh that would not withstand work outside. But for a family like hers, it does not matter. She would not have to work for a living, that's what servants are there for, or that's what her family thinks. But as Masa will say later in her life, after she has lived her life and at these moments when she wanted to impart a word or two of wisdom to her future daughter, Yoshiko, she will say, *Life is half and half. The stupid boy in my elementary school, the one whose family wasn't too rich, now owns a movie theater and is rich; the most beautiful girl in our class ended up losing her husband and became bitter. No matter who you are, we all get the same amount of happiness and grief in a lifetime.* Masa will lose her parents, all within several years of her childhood, and she will be raised by her nanny and her oldest brother. Before her parents die, she will have a younger brother they will name Gorō, the *fifth son,* and they will remain close until the end of their lives, hers sixty-nine years later, and his, twenty years following hers.

1917: The Restless World

The world is restless. Men demand changes here and there; anarchists are coming and so are socialists and communists. When the Russian aristocrats are dragged out of their beds in the middle of the night, when each and every one of them try to flee the country, some through the southern border to Germany and France, some westward through the tundra of Siberia to the Far East, their brother–sister royal families shudder in fear: *How dare the peasants,* they utter, t*hey don't know what's good for them. This is what happens when they start getting new ideas. This is what happens when they dare to go above their lot in life.* But they know that any day, any day now, they, too, can be dragged down to the ground, they, too, can disappear, and they tighten their grips. They become more and more vigilant in making sure that the people stay where they are.

April 1918: The Wrestler's Fever

No one knows where it started, only that it took the healthiest, the most vibrant ones of all, leaving behind orphans and widows. Not just an ordinary cold, but influenza, the Spanish flu, and we walk folded into ourselves, afraid to get near others, not sleeping with our children out of *fear* for the invisible, wearing masks, hiding, not hiding. A war is still going on in the distant land, to the west, though it may end any day now, or so people say. But here it is, we are dying quickly, in two days, in three days, all of a sudden, just like they say people die in war, but this is not war. The war is fought on the distant continent. The influenza is taking our lives. No one is safe, most of all children. Eiko and Saburō keep vigilant. Shirō, their first surviving son, must live. They will watch over him, they will do everything to save this son of theirs, the first one, the only one, and Eiko rubs her stomach. *There is another one here, let this child live as well.* By the end of 1920, 257,363 lives will be snuffed out, alive one day, dead the next, just like that, as if lives were as short as those of flies, of cicadas, as if the gods were indifferent to the suffering of the people. Some even say that the gods are punishing people for having turned their backs to what is proper, what is right, what it should be, for forgetting the gods that kept them safe and alive for many, many generations.

November 11, 1918: A Myth — The End of the War to End All Wars

A month before, in a small hamlet by the sea in Ibaraki, Shirō lies next to a baby: a younger brother, Susumu. Shirō does not know there is a war being fought somewhere in the world; his world is the farm and the river behind his house, and now, a younger brother. And later, his best ally, best friend, and closest brother. Both lives started during the War to End All Wars, and both will die during the Good War. But that is so far into the future. Now, here are a boy and his baby brother. Outside, there is a war that demands so little sacrifice, only 415 in total. A war that this little island country east of the Asian continent does not quite fight in, a war that asks so little. And when the war ends, all of Japan comes out and celebrates. *See, Japan has not lost a war, ever, in our twenty-five-hundred-year history.* A new myth has started.

1919: Life on the Farm

The morning starts even before the sun, because there is not enough time on the farm. There are not enough hours to go about; there are not enough bodies to sustain a farm this big. Even as a child, he is considered a labor force, a worker; there is no place for childhood on the farm. A child can feed the livestock. A child can fetch water from the well. Nobody lies idle on this farm, where there is not enough land to lend out to sharecroppers, which is not rich enough to hire help, but with land big enough to sustain, to trade. From the moment they are awake, their contract with the land starts. The livestock must be fed, people must be fed, the fire must be started. Water must be fetched to clean the dishes, and in the winter, the thick sheet of ice must be cracked open to get to the water underneath. Then to the land itself. The demanding earth. Everything is done with bodies: tilling the earth, leading the horse to plow, planting rice, one stalk at a time, with rows and rows of people bent at their backs planting, knee-deep in mud, jabbing the stalk in, then with a twist of the wrist, embedding the root deep into the mud. Then they can unbend their backs, look at the sky, and pray that the hurricane won't come to sweep everything away — that there's enough rain for a good harvest; that the river behind will not flood; that there's not enough sun to dry up the land. Pray for everything and nothing. A day on the land will make bodies crave for rest, for a dark dreamless sleep. The sleep is that deep, but it is not enough to rest the body.

1920: Start of the Rumbling Decade

All should have pointed toward the world without disturbance. The League of Nations started as a way to enforce peace. After all, the world still trembled from what had happened in the previous decade, with so many young lives interrupted, with the lands in Western Europe still marred by manmade trenches, with scorched land, with twenty-two million men who did survive the war haunted by dreams, with trembling they cannot control, in various stages of dismemberment both visible and invisible. *Shell shock,* they called it, but men aren't supposed to be afraid; they are not supposed to be affected. What they saw, these men, made language impotent, like it had made them impotent in life, made their relationships precarious, doubtful, and they retreat into their shells, shocked into their bodies, and they do not talk of what has happened. So instead, as if to forget, as if to revise, they dance, they drink moonshine, and listen to jazz and hold each other, even so briefly, even for a night, just to say, *See, we can still love, despite all that we have seen.* But that is rest of the world, and here, in Japan, the announcement that the Emperor is sick shocks the nation. A god has diabetes and sciatica. A god. Sick. We tremble. And as if to placate other gods, they open the Meiji Shrine in accordance to the auspicious direction decided by a priest, and Emperor Meiji and his wife are enshrined as gods. A quick escalation of two mere mortals to godhood. And here, at this moment, in Ōsu, Shirō and Susumu swim in the river behind their house, snaking amidst the boats, which they know they are not supposed to do, and Shirō does not know, or care, about what is happening in the world just yet, because he is a child, and he is at home in the small perimeter of his house, he and his brother, a world contained, perfect.

May 14, 1921: Birth of Kōkichi

Named *The One Who Serves*, Kōkichi comes into the world. Now, there are three sons. It is good. The war has been over for three years, the world all over is rejoicing about the new decade: no war, no famine. The era, Taishō. The birth of new ideas, of a new democracy. The Prince Regent, the first in the long imperial line, travels from the Far East to Europe, learning how to eat like a *civilized man* on board the ship, how to drink soup without slurping, how to eat bread, how to use a fork and knife without banging them against the plate. In Paris, after eating too many escargots, he takes a walk without a handful of servants, and, at one point, rides the Métro with only one officer in tow. But this is his first time riding public transportation, and he is giddy with the freedom he has never experienced, not at home, never at home. He does not know that a ticket must be bought, and it must be shown to a stationmaster when he leaves the station. He is scolded; this is the first time a common man has scolded him, and it tickles him. In England, he meets with King George VI, learns that kings and queens don't have to hide away in a mystical shroud. He is not yet a god, not yet the Heavenly Son; he is merely a young man, a Regent, learning that the world is big and that he is like any other. That he can be like any other, that he wants to be like any other. He will look back when he is an old man of seventy, when the world will have changed into such a different place, different air, and will say that this European trip is the most impressionable event in his life. But that is in the future; right now, the world is good, or so it seems. The world, at this point, does not see that the path to the next war is starting: a charismatic orator, Adolf Hitler, becomes the *führer* of the National Socialist German Workers' Party, a small, insignificant party in war-devastated Germany. People dream and dream of overturning their governments; in China, young Mao Zedong prepares to gather his followers in Shanghai. This is the time, still, because it is right after the War to End All Wars, because so many still remember how many boys came home dismembered, that peace is sought after. Officers of Japanese Imperial Army would take off their uniforms on their leaves, they are jeered at, sneered, in public, because soldiers, at this time, do not matter, because in the time of peace, there is no need for soldiers, whose job is to fight. For now, in a small hamlet by the river which tastes of the sea far away, where people travel from house to house on a boat, there is a family with three boys.

1922: *Moga, Mobo,* and All That Jazz

They pull pins and combs out of their hair. They chop off the symbolism of woman's beauty. They take off kimonos, the very cloth that constricted movements of their mothers and put on western dresses that give lightness to their steps. These girls take it all and make it their own: bob hair, short dress, stockings, painted eyebrows, and dark rouge. All the things their mothers would disapprove of. They strut down the streets of Tokyo, arm in arm, without chaperones; they flirt with boys, they ignore hisses from the older generation; they dance and dance until their feet hurt, but they can dance even then. They drink. They smoke. They hold on tight to boys, swaying their bodies languidly to music. A modern girl, a *moga*, laughs at people bound to tradition, she laughs at convention and modesty, despite the frowns, disregarding the social criticism, because she is all that, and more, or so she thinks.

11:58 a.m., September 1, 1923: The Earth Shaking, The World Ending

Birds stop midair. Horses grow restless in their stalls, pounding their hooves, slamming their bodies against the stalls. Signs have been around: rats running away in thousands, crabs crawling out of the sea, turning the roads into a moving sea of red, the moon bright red as if alit, the sky flashing with light without sound. The sky filled with dark serpentine clouds. *Look at the sun, it's so bright and the sky so strange,* people whisper. And suddenly, the earth falls from underfoot, then jerks. As if the world is suddenly a box, and gods are shaking it for fun. Books jump from the shelves; floors wave like the ocean during a typhoon; lunch dishes fly through the air; walls crack, then crumble. People trapped under ceilings, people pinned under ordinary objects that turn deadly with gravity. Her grandfather grabs Masa, throws her on his back, and runs toward the bamboo forest, where roots run deep, where roots go so deep into the earth that they go deeper than the womb of the earth. Houses collapse all around. Fire spreads, though the houses will not burn down, not in this farmland. Masa and her grandfather huddle in the green bamboo forest, grasping the stems, holding on to them even after the earth stops shaking. They stay there through the afternoon, when it boils up to 46 degrees Celsius, through the night, through the tremors and shakes, their hearts pounding every time the earth shakes. Masa looks to the southwestern sky. She sees the entire horizon burning up, as the imperial city so far away burns to the ground. She will see the same scene in March 1945 when Tokyo burned to the ground in the future war, but that is so far into the future, and she is not concerned about what will happen in twenty-two years.

September 3, 1923: Kill the Koreans

Tokyo burns furiously, lighting up the sky red at night. *The catfish god is angered by the decadence, by the sad state of the nation,* someone says, and others nod. See how the earth shook, see how fire spread like hands of the angry gods, destroying all. There is no god to keep the catfish down; Kashima is turning a blind eye. That's what they said during the Edo period, and it's still true. The nation needs to be changed; a new society needs to be born out of the ruins. With the prime minister dead, with the Emperor standing behind the chrysanthemum veil, who is there to protect us? This is a sign, the rumor starts. This is the sign that we need to change the world. With newspaper offices shut down, hysteria spreads. *Koreans are rioting, they are poisoning the well water. Koreans are raping our women and setting houses on fire.* How ungrateful, those Koreans. They are our colony, we've rescued them from backwardness, we're leading them toward modernization. We've made them a part of our holy and sacred territory, but look how they sulk. Look how they look at us, always thinking of ways to cheat us out of our hard-earned money. After all that we've done for them, this is how they repay us. Catch those Koreans. We know how to find them: they can't pronounce *ga-gi-gu-ge-go, jūgo en,* or fifteen yen. Before they have the chance to loot, before they have the chance to poison more of our water, kill those bastards. Gods are angry because of them. Kill them.

September 25, 1923: The Imperial City in Ruin

Tokyo remains dead. The Prince Regent has come from behind the chrysanthemum veil and walked amongst us twice. He has walked amidst the ruins, He has opened up His treasure chest and houses us with His own personal money. How He pitied us, while He held His own personal grief hidden in his own heart. He has put aside His own impending marriage, pushed it to the following year so that the nation can recover from this cathartic event. People are still trapped dead, the smell of Tokyo fouled with the stench of decaying flesh. Koreans are massacred; so are the deaf and mute. 140,000 people died from the earthquake; somewhere between 3000 to 6000 Koreans and disabled people murdered. The Imperial City is under martial law; no one can walk outside after sun has set. The living must keep going. We must rebuild the Imperial Metropolis. We must rebuild the nation in ruin.

December 27, 1923: The Attempted Assassination of the Prince Regent

The car carrying the Prince Regent pulls out of his sacred palace, and we bow deeply as we line the street, so as not to pollute him by showing our faces to him. We bow, we bow deeply, and the car keeps driving toward Toranomon, but there's no end to us lining up the street, bowing deeply and, as if by mistake, a young man breaks away from the sea of bowed bodies, a young man takes hurried steps toward the slowly moving car and raises his cane and takes aim. A loud crackling sound. A gun hidden in a cane. We don't know what has happened, just for a fraction of a second, until the car drives away quickly with a shattered window and the young man begins to run, shouting, *Long Live the Revolution,* and we rush, we rush toward the man, we rush as a mass, we move as an entire wave, *kill him, kill him,* we pull on his jacket, we claw at his eyes, at his face, a policeman has rushed in to protect him and we punch the police, *kill him, kill him.* And we would have, if it weren't for that policeman getting in middle. The newspapers, for days and months following, scream of the traitor, of how his father has resigned from the Senate, how this traitor, Daisuke Nanba, had written letters to his former friends to break off their friendship before his act, and later, how his father refused his son's body after the execution. We will think it just punishment. He has tried to do the unthinkable; he paid for it. And we will think nothing when the newspaper writes of how his father locked the gate of his house and starved himself to death. After all, this is all a fitting fate for a traitor who dared to attempt the improbable, and the family must pay for the sin of the son.

January 26, 1924: Marriage of the Prince Regent to His Distant Cousin, Princess Nagako

The Imperial City is still in ruins: parks filled with homeless families, streets overrun with human excrement and garbage. It is a hard winter for people sleeping outside, and it is unusually cold this winter. But here it is: an auspicious wedding between the Prince Regent and his distant cousin. He is a man now, a family man; he is already a prince regent, and now, he is married. Now, we will wait for a boy. And all will be well when there is a boy prince to carry on the family name and the uninterrupted twenty-six-hundred-year reign by one family.

1925: Shirō, a Bright Child

Shirō is now nine years old. And for a nine-year-old boy, the world is his. Everything is possible, and even the dreams he dares to dream are as real, as tangible, as the ground he walks on. He walks to school with Susumu, a branch school where all the grades sit in one class with one teacher going around helping each pupil, some eager, some lazy, some indifferent, all children of farms in the neighborhood and most would not need anything beyond basic writing, reading, and arithmetic, just enough to fill out a tax form, enough to fill out forms to take out more loans from the neighborhood fund or banks, enough to figure out how much money is needed for next year's seeds. But Shirō is different, or so he has been told. He is the first son of a big farmer and he needs to make his family proud, to make sure he does not taint the honor of the Shimura name. And because he is earnest, because that is the curse and the blessing he will carry for the rest of his life, he studies hard, he does not allow himself less than the perfect score. And all he has learned, he takes home to his grandmother and he teaches her the day's lesson because she is a woman of the previous generation whose worth was measured by what she could do around the home, not by her reading skills or knowing facts from history. She is illiterate; for her, the letters on the page will remain incomprehensible for the rest of her life, and it is Shirō's job to read her newspapers every day, it is his job to recite all that he has learned, it is his job to write out her name, one stroke at a time, so that she will smile and tell him what a smart boy he is. He takes his job seriously;

he is the first son. It will be his job to take care of the family when he grows up, it will be his burden, like a horse tilling the earth, like geese that must migrate, winging the vast distance from the north across the ocean to where he is.

December 25, 1926: The New Era Starts, the World Darkens

The Taishō Emperor, the weak one, the sick one, has been unable to go one, unable to stand up, unable to eat, while his son has been growing stronger, older, now married. Taishō Emperor lived as a man, walking amidst men, and died a child he never was, holding the hand of his biological mother, a concubine of his father, whom he has not seen for a very long time, not since he was separated from her and given up to be raised by others, then adopted by the Empress. The entire nation goes into a deep mourning, though we have known that he has been weak, he has been sick, he has been gradually disappearing from the public life. Now, his son, the Prince Regent, is the 124th Emperor. The family of the direct descendent of Amaterasu Ōmikami, the Sun Goddess. His era is named Shōwa, and he will reign for sixty-three years. He will reign the first twenty years wearing the military uniform, and he will reign the last forty-three years wearing worn suites and a fedora hat. He will rise as a god and he will die as a man.

April 1928: One of the Two Bicycles in the Hamlet

At the time when one bicycle cost more than the monthly salary of a recently graduated banker, the family owns two bicycles — one for Shirō to ride to the middle school, and the other for Susumu, who will soon follow him. At the time when only 240,000 out of 1.6 million children go to middle school, when a farmer's eldest son is often told, *You don't need any education, you only need to know how to write your name,* Shirō has passed the entrance exam with high marks; when tennis was played

Fig. 1.2. A page from Shirō's sketchbook.

only by the rich and the aristocratic, Shirō has joined the tennis team and is making progress that astonishes his teammates. He is the junior kendo champion, as fitting for the child living right in between the god of Kashima—the god that suppresses the earthquake—and the Narita neighborhood known for its fierce warriors, though the time is so far away from the Edo period, when only the selected were warriors. Now, in this era, every man is allowed to be a warrior, only if the nation deems them fit. Like all boys, Shirō, without knowledge, is being monitored by the town hall; each and every boy's records, both personal and public, are kept in folders in the backroom of the office, and all his achievements, all his grades and health—or sickness—are meticulous noted in small handwriting to be opened when he is twenty, at the time of his conscription, but that is still eight years into the future. Let the boy be who he is for now. Let the boy ride through the rice fields under the stark blue sky with his school bag strapped on his back, in awe at his own speed, of his own prowess, cutting the wind in half.

March 1930: The Showa Market Crash

The wave, which started on October 24 last year, has finally reached this shore. In New York, it rained papers rendered meaningless by the Crash, it rained men in suits, and the land filled with men sitting, waiting, squatting, waiting, standing in lines, for jobs, for money, for food while their wives waited at home, worrying about the probable future awaiting them. And here, in Japan, the same begins. But instead of men jumping from buildings in cities, three million men lose their jobs and go back home to the countryside; *geishas* lose their jobs because their patrons go bankrupt, and money becomes tighter than the fist holding it. The focus shifts to the farmland, especially northern Japan. Daughters see their famished siblings, their bellies bloated and their limbs turning into branches, hard, lusterless, from hunger. They know what they have to do; daughters are extra mouths, they are unproductive, they will not carry on the family name. They are, as they have always been, unnecessary. It's not only the parents selling them; it's them, the daughters, selling themselves. Seventeen yen for four years of service, money they will not see. They go into factories; they go into maid services. And in cities, where money is abundant for a handful, under the brilliant colors — red, gold, yellow — soon, soon enough, they will realize that they can make more money without working like slaves, in cafés, as prostitutes, as overseas prostitutes, *karayuki-san,* because away from home in cities and in foreign lands, they can be anyone, they can remold their past and present at their will; and when it is time for them to go home, they can erase their tainted pasts and live the lives they had left off. Without education, without the protection of their family, without skills, what else can they sell but their bodies?

IMAGINARY DEATH

Fig. 1.3. A page from Shirō's sketchbook.

April 1931: Shirō Attending the Agricultural Vocational School

Instead of going to a university in the Imperial City, as he had hoped, instead of making his dreams realized because he believed anything is possible, Shirō goes to the Agricultural Vocational School, just like his parents wish. He tells himself that this is what a first son of a farmer is supposed to do, that he has had more education than most boys he grew up with, that this way, he can learn all he can about the modern technology of agriculture and make their — *his* in the future — farm as bountiful and productive as possible. Of course, he wants to go to high school. Of course he wants to go to university. But he also knows that he is the eldest son, that his dream — if he dares to dream — must not be just his, but his family's. So, every morning, he wakes up and quickly eats breakfast, jumps on the bicycle and pedals through the rice paddy, still before planting, with the water in front of him reflecting the early sunlight, straight toward the big river, across the Jingu Bridge, which people say has human sacrifices chained to every pillar underwater. A path he is familiar with, only nine kilometers one way. The air smells of the ocean he cannot yet see, but will once he crosses the river — or is it an inlet? Cattails, tall; marshes. He pulls himself high on the saddle, almost standing on his pedals, and crosses the river. And after that, all he has to do is veer right, and the school will be there, at the end of the gentle curve, and he will get off the bicycle, he will bow to the small altar containing the sacred pictures of the Emperor and the Empress, he will say hello to his friends and sit in the class to learn geology, chemistry, English; all the things that he may or may not need to be a good farmer. On a good day, he can be here in thirty minutes; on a bad day, forty-five minutes. Today is a good day; he has broken his own record.

September 8, 1931: The Manchukuo Incident

The tension in northern China has been building, like a volcano about to erupt; the rumbling and trembling of the region felt in all the eight corners of the earth. Japanese immigrants are driven out of their houses by Chinese rebels who used to be farmers living on the very land the Japanese have taken; Koreans, driven out of their homes when the Japanese came to their land, moved into China, and they, too were attacked by Chinese farmers who wanted their land back. The Nationalists run around causing havoc, *Japan has no right to be here.* The Communists fight while telling the people, *The cause of your misfortune is the Japanese here.* We back in Japan do not know all this: we know only that Chinese are massacring Japanese for no reason. We have other things to worry about; we need to find jobs, we need to find money, we need to feed ourselves. The economy has spiraled down to nothing so that farmers down in the south, up north, cannot help but to sell more girls. The cycle of hatred whirls in the region until something had to give. And something does give. The Kwantung Army plots. It plots to create a start, the reason, and it does. On this day, the railroad line in northern China explodes, and the soldiers of Kwantung Army spread like fans, taking one city, one army at a time, spreading, claiming as their own, all in the name of the Emperor–god, all in the name of the Japan Empire, although the Emperor–god does not know

IMAGINARY DEATH

Fig. 1.4. A page from Shirō's sketchbook.

what has been occurring in the minds of his men in Kwantung Army. And six months later, the explosion spreads to the southern China, to Shanghai and later will spread throughout all of China.

January 8, 1932: Assassination Attempt of the Emperor–God at Sakurada-mon

As the carriage with the sacred chrysanthemum insignia passes us by, as it is about to enter the Imperial Palace, something flies through the air. A hand grenade. And it explodes. Horses carrying the Emperor's Guards bolt; the carriage horses are whipped into the palace, to protect the Emperor–god. Another carriage sits still amidst the rubble, amidst the sudden chaos, with its left rear wheel splintered. People run this way, that way, screams and yells mobbing the place. Horses, still attached to the dismantled carriage, lie on their sides, struggling, spluttering blood with each movement, the whites of their eyes in panic, in pain. *That's him, that's him,* the Emperor's Guards yell, the special policemen yell, no, no, a man is caught, and someone else yells, *He didn't do it, I did it, I did it,* a young man with strange accent, a Korean man, and the guards rush to him, mob him, hold him down to the ground, pin him. A Korean independence activist, Lee Bong-chang; an assassin, a man with a mission who didn't know which carriage the Emperor–god would be in. Ten months later, he will be executed for high treason. Koreans in Korea, reading a tiny article in the corner of the newspaper, rejoice, though they keep their faces blank. They know there are eyes everywhere; anyone can turn them in to the Japanese police. A hero to one, a traitor to another.

February 22, 1932: A Myth — The Birth of the Three Bombing Warrior Gods

Here are three gallant warriors who opened a path to other infantry soldiers, who strapped themselves to a bomb, threw themselves against the wire fence four meters high, and blew themselves up, one newspaper headline screams. Another yells, *These three bombing warrior gods' act of ultimate sacrifice is the true reflection of the god-like race of men.* Three privates, on this day, hurled themselves against the enemy line with a bomb in their arms, knowing that they would die in doing so, and blew themselves up to open a way for their brother-soldiers to attack the enemy. Newspapers say that they went running yelling, *Long Live the Emperor.* See how they forged ahead, knowing that they would die. Newspapers say that the Emperor has been notified of their valor. Oh how proud we are, we tell ourselves, this is indeed a just war, a holy war, see how men will happily offer their lives to the cause. And these men, why, these men are just like us, sons of miners and farmers and a carpenter's brother. Just like us. Brother-soldiers in their unit tell no one of what they really saw: *These men who were named the warrior gods — they were supposed to light the Bangalore torpedoes, run to the wire fence, and retreat as quickly as possible. But on the way there, one of the men tripped or was shot, and they fumbled so three of them came running back without bombing the fence. So our sergeant yelled at them,* What the hell are you doing, go back, go back and finish it for the Emperor, for the country! *So they had no choice but to go back. The bomb blew up in their arms just as they reached the fence. They probably knew that they were going to die — either by getting shot for disobeying an order or blowing up. It really was pitiful.* Not one of the brother-soldiers said how impossible a mission it was for three men to run for several dozen meters of rough terrain in the dark, carrying a twenty-kilo, four-meter bamboo stuffed with explosives; not one of them said how impossible it was to run several dozen meters with a lit fuse, shorter than usual, under heavy enemy fire. It was suicidal, it was a bad command. But for now, the newspapers mold their stories into a selfless act, a representation of the just war, the justification of a war being fought in a faraway land. These men, now martyrs, have been raised two ranks after their deaths. Boys all over Japan strap themselves with pretend-bombs and play the warrior gods; girls daydream of adorning themselves in white, nursing wounded soldiers back to health. All of Japan celebrates the birth of new warrior gods.

1932: The Emperor Is Rising to Godhood

He has always been a heavenly son, because he sits behind a mystical shroud, because no one is supposed to see him, as befits a high priest. But there is a movement among officers of the military to raise him to the militant strong Emperor–god His grandfather was, unlike the Taishō Emperor, the weak one, the one who was rumored to be *slow,* the one who stopped his speech to roll up the document and use it as a telescope. But no one talked about the dead Emperor's condition, and now no one must talk about the current one's mortality. Whatever may be the case, *we need a stronger ruler, civil servants can't be trusted; the politicians can't be trusted, we need a ruler who can override these selfish people, who can make Japan proud, fight against Western decadence, a ruler who can lead us back to the Japanese spirit.* The Emperor is slowly moving toward godhood, He sits on the throne, carried aloft by the militants, by the newspapers, by the people as they do during a festival, and He cannot get off.

May 15, 1932: The Assassination of the Prime Minister by the Imperial Naval and Army Officers

Things didn't go exactly the way they had imagined; it was supposed to be an easy coup, the kind that the samurais did in the golden age of warriors — men armed, ready to forsake their lives, just to change the world. It wasn't supposed to be an assassination; it's too base a word to describe what they are doing. This is a sacred mission to cleanse the society of vermins. The plan: teams to divide into several groups: to assassinate the Prime Minister, the one who is against the expansion into China; then to the Bank of Japan; the Metropolitan Police Department, the Bank of Mitsubishi; then at 19:00, to the electricity plants to black out the imperial capital. But the old man, the Prime Minister, just would not make things easy. The plan: Team A was to meet at the Yasukuni Shrine at 17:00, then to go to the Prime Minister's residence by 17:30 and to shoot him as soon as they find him. But the old man just would not let them — he invites them all to the living room, he scolds the officers, *At least have a decency to take off your shoes inside.* He opens his mouth, he is about to say something — that's when another team bursts in, and the old man says, *Wait, wait, let's talk, let us talk,* but a yell, *No need to talk, shoot, shoot,* an officer shoots him in his stomach; another takes his gun, aims it at the wrinkled right temple, and shoots. Even then, the old man is still alive, still lucid, and as his maid comes in and the officers rush out, they hear him telling the maid, *Get those young men back here, I'm not done yet.* Things just don't go at all. The old man. And where was Charlie Chaplin who was supposed to be there? Only if they had killed Chaplin as well, if they had killed the American comedian, they could've shown to America, to the rest of the world.... They are trying to change the world, they are trying in vain to right the wrong; as they had written in fliers and thrown them out of the running car,

Fig. 1.5. Shirō as a young man, 1933.

Japanese people!

Look at the reality of what is happening to our motherland, Japan

Where in politics, in diplomacy, in economy, in education, in ideology, in military, can we find the authentic Imperial Japan.

The politics which is blind to all except for the interest with zaibatsu to squeeze out the last drop of blood from the masses; the tyrannical authorities suppressing the people while protecting themselves; weak diplomacy and decadent education and rotten military and worsening ideology and the suffering farmer and working class and....

People! Arm yourselves and rise! The only way to save this motherland of ours: with direct action.

Japanese people!

In the name of the Emperor, butcher his treacherous subjects!

Kill the political parties, kill the zaibatsu!

The world is rotten, it really is, we think; look at these officers who put their lives in line to save the Emperor from his rotten subjects, look at how they all stand proudly, although the newspapers call them *butchers* and *traitors*. We say to ourselves, *How can we read the letter without tearing up, look how they were trying to save us,* and we sign petitions asking for their leniency, and they do get it. No one is executed. No harm is done.

9:15 a.m., December 16, 1932: Lions Roaring, Women Falling from the Building

The fire started out on the fourth floor, from the Christmas tree overburdened with decorations, and it spread quickly upward. One of the biggest department stores in Tokyo, Shiraki-ya, burns in the middle of Tokyo, and people watch it from the ground. Inside, customers and sellers clamor down the staircase, as fire snakes upward, upward, engulfing the fifth, sixth, seventh, and eighth floor until it reaches the rooftop, where two lions roar in panic, but no one wants to help them. Instead, they are fleeing, fleeing, pushing each other down the steps, just to get to the ground floor, just to get down to the street. A man falls to the ground when he slips from the bolt of silk kimono he had unrolled to use as an emergency ladder; women fall, one by one, letting go of the rope because they cannot bear for gawkers on the street to see their naked buttocks as they slide down the rope. The first fire. The newspaper sings, *These women died rather than be immodest,* and women everywhere begin to wear pairs of panties, for modesty, for chastity, because this is what Japanese women do. They tell themselves, *we would rather die than live in shame.* Another small myth is added to the national myth.

February 12, 1933: The Vague Uneasiness

"Do not let a single woman or women in pairs go see the volcano mouth by themselves. They must be accompanied by a guide at all times," warns the police newspaper. The year, 1933. It all started when two young girls traveled to Ōshima on a ship, hiked up Mihara Mountain, and one of them jumped into the volcano as another stood as a witness. She came down by herself; but this was not her first. This was her second pilgrimage as a witness only a month ago. This decade, which started out with the crash of the stock market on a distant shore, its wave hitting Japan a year later, is filled with girls selling themselves in famished villages and dotted with college-educated men without jobs and children so malnourished that they died as quickly as they were birthed. A dark era: a bad economy, a vision so unfocused, their lives without concrete meaning. The vague uneasiness. Unmoored lives. The era of eroticism, the grotesque, and terrorism. The economy picked up because of the *Manchurian Incident,* the nation filled with men in the army, with factories running nonstop for war production. But something is amiss. An uneasiness. A shadow of something dark to come. This year will find only 944 men and women jumping into this very volcano, and overall 14,805 people will kill themselves this year, with 1,936 the highest number of suicides until 1937, with the outbreak of the *China Incident,* when men will not have a choice in ending their lives but submit to the fate coerced by the State and the Emperor–god.

How to Avoid Being Drafted

Some do not want to be drafted, no. They do not believe in serving in the army, wasting two years of their prime time; some are the only wage-earners, the main breadwinners, they cannot afford to be absent from home, not when there's a farm to run, not when there is an elderly mother whose only means of survival is her son; some, out of principle, but they will not say it; some do not care if they become men in the eyes of the nation or not. If they can afford it, they can go to universities or immigrate to America, to Manchuria, to somewhere far away from the mainland. Women go to shrines to pray to local gods they grew up praying to so that their husbands and sons will not be taken; women go to seers to get talismans so that their men are not chosen. But they can do only so much. So here it is, this is what sons can do: they starve themselves so that they can be underweight, they can drink gallons of soy sauce to initiate heart palpitations, they can insert wax and shell fragments and egg yolk into the ear canal to pretend they are hard of hearing, they can cut off a finger or two, they can pierce their genitals with needles to create internal bleeding, some even inject paraffin in their jaws to create a deformity. All of this, in order to avoid being conscripted. All of this, a quiet resistance against the system that demands that it is the obligation of men to serve its nation, but they also know that if they run away, they will not have the identification card that the ration system demands, that their families will be ostracized from the community, and that there is no way for them to survive in a nation that has followed them from the moment of their birth.

IMAGINARY DEATH

Fig. 1.6. Masa (right) as a student at Mito Women's College.

1933: Two Photos

One photo of Masa during her time in a Women's College; a privileged young woman's life. She is living in a boarding school, first time away from home. This photo is of her in a studio: she is standing next to a girl who is sitting down, she is wearing her school uniform — an overcoat and kimono-pants — as is her unnamed friend, their hair in loose buns as is the style of this time. Her gaze seems to be looking at something beyond the photographer, her gaze not quite settling in center, but to something not far away; she is not smiling, but then, neither are most photos from this time period. It is not the face of a dreamer, but a realist, someone who understands that after her time in a Woman's College, she will marry the one chosen for her, because that is what girls of a certain age do; younger in the rural area, older in the cities, but all women are expected to, or they are considered nothing. And having gone to a Woman's College, Masa's chance of getting married is limited in a small town like this. Now that she has education, her hands are not meant to be dirtied; her hands are meant to sew, to touch, but not to be used for labor. And how many men can afford servants in this town? How many men have education enough to match hers? Here is another photo: Shirō and Susumu, two brothers in near-matching plaid wool kimonos, and caps with visors on their heads, the symbol of a student. They are thickly built; Susumu sits and Shirō stands. Shirō's hand is clutched, and there is a faint smile hovering, but not quite. Susumu, his younger brother, is gazing the opposite direction. Shirō has been shooting with live ammunition as a part of the basic drill at the Agricultural School, but then, it is a part of the curriculum: to cultivate soldiers before they are twenty, to grade them, to judge them, to deem who is worthy and who is not. Just like the basic training he will be experiencing when he is conscripted, he runs with a heavy pack on his back every day; he crawls, with others, through mud and rice paddies with guns in the darkest night, commanded not to make a sound, the retired drill sergeants shouting commands, taking note, and next to Shirō's name, highest marks in everything. The note will later be forwarded to the town hall as part of Shirō's record. Susumu, on the other hand, is defiant; he would rather play soccer than have to submit to rules, to fanatic dogmas. His younger sister would later say, *He was hated by the drill sergeant, that's why he never advanced in the army when he should have; after all, he went to university. He should have been promoted to an officer track after the three months of basic training.* Susumu's record must have reflected all this. The economy is booming because of the war in China. The nation is becoming more and more militaristic, when they must submit to the call of the nation that they are living under a *time of emergency.* But these battles are fought in a faraway land; the economy is good, but it does not affect individual farmers. For now, let these young people dream their dreams: they have yet to meet.

February 24, 1933: The Orphan of the World

The verdict at the League of Nations: forty-two in agreement, with one against, and twelve absences. He had done his best to persuade the rest of the world to see that Manchukuo is a legitimate nation; he gave a speech for an hour and twenty minutes in his fluent English he had learned as a young man in America, the English of the bible, and he speaks in biblical terms, justifying Japan's action in Manchukuo, that Japan will not leave Manchukuo. But Japan is the only nation to vote against the infamous Lytton Report, the report that states that Manchukuo is not a nation, not a real government but only a puppet state of the Japanese empire. How can he make *them* see that Manchukuo is the *lifeline for Japan's survival,* that Manchukuo offers new lives to second-and-third sons of farmers who cannot inherit any land, that it is a promised land to an already overpopulated motherland, that it offers resources that otherwise they cannot be get? The diplomat, *usually typifying the placid oriental diplomat,* becomes more and more passionate as the prepared speech goes on, and at last, he throws it away, instead shouting, *Japan will oppose any attempt at international control of Manchuria.*

It does not mean that we defy you, because Manchuria belongs to us by right. Read your history. We recovered Manchuria from Russia. We made it what it is today. He sees that no one agrees; he looks at each and every face of the diplomats, and his words are not being heard. He continues, referring to the report that Manchukuo be internationally governed: *Would the American people agree to such control of the Panama Canal Zone; would the British permit it over Egypt?* The vote is taken. Japan is the only one who sees the rightness of the action. The Japanese diplomat, in his finest, gets up again. He takes out the prepared speech, the one he had hoped he did not have to use, and begins to read, *The Japanese government now find themselves compelled to conclude that Japan and other members of the league entertain different views on the manner to achieve peace in the Far East, and the Japanese government feel they have now reached the limit of their endeavors to co-operate with the League with regard to Sino-Japanese differences. The Japanese government will, however, make their utmost efforts for the establishment of peace in the Far East and the maintenance and strengthening of cordial relations with other powers. I need hardly add that the Japanese government will persist in their desire to contribute to human welfare, and will continue their policy of co-operating in all sincerity in the work dedicated to world peace.* With that, he utters, *Sainara,* a casual way of saying, "We are not coming back," takes off his owlish glasses, and stomps out of the League of Nations, proud, angry, with his back straight. He comes home, hailed as a hero, as a representative of a country that can say no to the Western nations, the only country in the Far East that can stand up to the bullying of the West. Now, Japan is an orphan; or is she a runaway, forsaking brotherhood of nations to be a renegade?

6:50 a.m., December 23, 1933: The Birth of the New Crown Prince

All over Tokyo, sirens blare for a minute straight, and on cue, the entire nation comes out to celebrate. A child is born, like any other child, but this is a special one, blessed by the gods, the one so awaited by the entire nation. How we waited for this day, and every time the Mother–Empress became pregnant, we held our breaths, and every time we found it was a princess, we sighed in disappointment. But now, a boy, a boy to carry on the family, the boy to carry on our nation. The new crown prince is born. We are happy. The gods are pleased.

1934: The Myth of War Widows, or What It Means to be the Mother of the Emperor's Children

Widows must not cry. Like the wives of samurai — the fabled warriors of the past who feared shame and disloyalty more than their own death — women must remain calm, unapproachable, statue-like. Their husbands did not belong to them; they belonged to the State in life, and in their deaths, they will protect the Emperor–god and the nation is kept safe. Women, the magazine articles command, keep your back straight. Tell the children, *Your father died bravely for the Emperor, you mustn't cry, he's now at the Yasukuni Shrine, and when you grow up, you must also die for your country.* The urns will arrive ceremoniously, solemnly, each urn carried gently by soldiers, one soldier for each dead. There will be a ceremony held by the village. The houses of the dead will be called the *revered home*. The widow will raise the children on her own, based on the compensation paid by the government. She will not cry. She will keep her back straight; she will raise her children to be the most loyal soldiers of the Emperor. *That's what he would have wanted,* she tells everyone. *He would have wanted his children to serve and die for the Emperor–god.* This is what she is told, her gesture of grief informed by magazine articles and newspapers, her private mourning molded by the imagined mode of samurai's wives. She does not know that the samurai code is a construct of a later day, during the Edo period, when there were no wars to be fought. Right now, there are so few war dead, and women, looking at the pictures of these stoic widows, the wives of *chosen ones,* true women because they were married to the true men, file away the images for the future.

1935: The Gradual Molding of a Boy into a Man

It is not the physical exam at conscription that changes a boy into a man — although he must undergo the humiliation of standing naked, blushing with shame as a doctor's thick finger roughly probes every orifice. It is not the moment when the result is shouted in front of the others, all naked, all standing taut as they have seen the veterans in their village do. Fit for conscription, unfit for conscription, their manhood declared by an Emperor-god's servants. It does not happen when he enters the barrack gate in his finery or when he strips off his civilian clothes, identity, status, and past, remolding himself into a uniformed man devoid of individual history, his only name his family name. It is not all the obedience and submission he gets punched into him every day for three months during basic training, until he would do anything he was commanded to do, no matter how unreasonable, and shame comes easily until he is reduced to a fragmented body that can only parrot the collective language of nationalism. It is not the first kill on the battlefield. The making of a soldier does not occur during the two years of military conscription. It occurs so slowly, so gradually, even before he stands naked for the physical exam at age twenty. It starts while he is still a boy in school: even the child who doesn't know how to read yet knows that the most important person in his life is not his parents, not his local gods, not his ancestors as he was taught at home, but the Emperor-god upon whose face no one is allowed to gaze. By the time he is twenty, a boy is ready for conscription. He is ready to appropriate the manhood so cherished, so prized in the eyes of the nation. He can say, *serve the country, long live the Emperor,* as if he is saying *good morning, good night.* As if these words are a part of his body.

January 1936: Marriage of Masa and Shirō

The bride sits on the boat coming down the river. The bride in the finest black kimono, carrying the family insignia, comes down the river amongst the fleet of boats carrying all of her dowry: robes and robes of kimono for all occasions, the best kind that money can buy, chests made out of the most expensive paulownia woods, barrels of rice. The bride comes down the river in the finest procession this hermitage has seen in many, many years. The neighbors stand by the riverbanks on both sides, in awe of the pale skin of the bride, of how many boats it takes to carry all the dowry she is bringing with her, that she's rumored to have graduated from the Women's College in Mito, that she comes from the wealthy and old family. Look at her tortoise combs and pins in her hair, they whisper, as she passed by them; you can't buy something like that, it must cost a fortune. Shirō stands by his house by the river, waiting at the dock. He has known that this day would eventually come — there's been talk about his marriage for a while now. It was a matter of which one his family — his father — would choose for him: this girl or that girl; several pictures had come to the family as soon as he was of age. And perhaps with the incident in China, the army mobilizing to China to fight against the Chinese, to protect the Chinese from the tyranny of the Nationalists or the Communists, it didn't matter which, it may have been his father's subtle and unspoken wish for Shirō to fail the conscription exam. After all, *they* never took a married man; *they* never took an eldest son; at least that's how it was when Saburō was a young man. Shirō has seen her picture: she was not a beauty, not like the girl he had secretly liked when he was still a student, the girl whose name he kept doodling in his sketchbook. The portrait of a girl he had drawn in the sketchbook. But that is not the girl he is marrying. What he knows of the bride-to-be: Masa Ōkawa. Went to a Women's College. From one of the oldest families, one of the richest families in the neighboring village. A distant relative by marriage: Saburō's niece — Shirō's cousin — married into the Ōkawa family. A marriage of this kind would have never happened, this marriage between families of such different classes, however subtle it may be amongst the farming family. Her family owned farmland, but they never did have to toil with their bodies; his family owned farmland, but it is one of the branched families of the Shimura Clan, not the main line, and they do have to work with their backs bent, each and every one of them. But somehow, the Ōkawa patriarch — her brother — agreed. And Saburō rejoiced and Shirō blushed. It may not have been what he had secretly hoped for when he was younger, that the girl he marries would be the girl he loves from the bottom of his heart, although he also knew that in a family like his, something like this seldom happened, that romantic love only happened in cities, when there was no burden of the family name for recent migrants. But he has seen this: that an arranged marriage like this can work, that liking can turn into lifelong respect and love, so much so that when an old couple walk next to each other, they resemble each other from years and years of shared history together. He has also seen couples being miserable, years and years of bitterness creating a distance that cannot be overcome, no matter how many children they have, no matter how many years they work side by side. His marriage will be different, he thinks. His marriage will be the book of the old tales from *Tales from Now and Past,* where the couples long for each other so much that the woman, even after she dies, waits as a ghost for her husband to come home; she just could not go to the otherworld without seeing him, just once, and making love to him. His marriage — their marriage — will be a happy one. This is what he thinks as he stands there in his finest kimono adorned with the family insignia of an ear of rice, as the fleet of riverboats approaches him, as his bride comes toward him.

February 26, 1936: The Rebellion of the Emperor's Guards

It is snowing; it kept snowing, but not fast enough to erase the footsteps of thousands or so armed soldiers, both the Emperor's Guards and the Imperial Army, marching toward their destinations: the homes of the Prime Minister, the Minister of Finance, the High Secretary of His Majesty, His Majesty's Lord-in-Waiting, and others, as well as capturing the office of the *Asahi Shinbun* newspaper and barricading the Nagata-chō neighborhood. The officers of these regiments have mobilized their men to do one thing: to eradicate these errant subjects who use His Majesty the Emperor to do what they want to do. They must return Japan to the Emperor, to let the Emperor lead Japan to whatever fate awaits it, and they reason, *as long as these old men stay in power, the Emperor will not be able to lead like an Emperor is supposed to.* The reason why the empire has come so far, on par with the West, is because of the Meiji Emperor, and all that has gone wrong is because of His servants who made the wrong choices; it is because of these old men that His Majesty does not see what is happening to Japan — families starving to death; girls selling themselves to prostitution. The rotten political system in league with the *zaibatsu* system. The only way to have an authentic Japan is for the Emperor to return to His proper place: to rule. They know this is what His Majesty wants; they *believe* it. It is early in the morning; they steal through the snow to their destinations. The Prime Minister's house is the first to be attacked, but a maid hides him in the closet, and instead his brother-in-law, who so strangely looks like him, is shot dead. The Minister of Finance: shot dead. The High Secretary of His Majesty: shot down with machine guns; his wife tries to stop them by throwing herself in front of her husband and is shot down, while his nine-year-old daughter hidden behind the furniture witnesses all of this. His Majesty's Lord-in-Waiting, the most trusted by His Majesty: shot and fatally wounded, but when his wife — the very woman who had raised the Emperor–god for a decade — begs them not to kill him, an officer salutes and leaves. One by one, old men are taken out and with each man dead, the future seems more and more possible. Or so they thought, until the Emperor is woken up with the news: His subjects are being massacred by renegade officers. He cannot believe it, *How dare they,* he shouts, *how dare they, if you do not stop them, I will personally lead the army to suppress them.* Three days of rebellion, and under the command of the Holy rescript, men under the officers' command put down their arms and walk away. They were just following orders, not knowing that they were contributing to a rebellion, but later most of them would be sent to the foremost battlefronts, most of them would perish in China. This very moment is the triumph of the militarists. This is the moment in history when the Emperor chooses that he is indeed only a constitutional monarch, not an absolute monarch like those rebellious officers wanted Him to be.

May 1936: Shirō, at the Conscription Examination

He stands naked in a way he has never stood amongst others. He knew this was coming; the notice came at the end of January. But he did not know that from the moment he was born, with every step of the way in his life, the village office kept record: his grades, tennis competition results, sword contest championships, height and appearance, achievements, personality, intelligence, skills, his family's income and property size, all the things as if there is no privacy, and there is none. A child is not of the family but of the state. And every child is one of the Emperor's children. Now, he stands naked, in attention, his back straight. His boyhood friends stand around him, the way they have seen veterans do, their backs so straight they almost keel backwards but do not. One by one, they are called to the table. Shirō is called. A neighborhood doctor tests his hearing. Eyesight. Weight. Height. And for venereal disease: the old doctor roughly grabs his penis, squeezes it hard again and again to see if any puss oozes out. Shirō has been warned: in the manual sent before this examination, it warned against buying a prostitute, how catching a venereal disease is unpatriotic, how it would destroy you, your family, and the future of Japan. Then the old doctor tells him to get down on his fours on the cold floor, just like a dog. Without saying anything, Shirō suddenly feels a thick finger jammed into his anus. He flinches. The finger probes, moves about the interior wall, and once it is done, an officer tells him to get back on his feet and stand in line. Shirō goes back to where the boys are, and they wait in attention, still naked as the doctors and officers confer. And one by one, they yell out: So-and-So, *unqualified*. The boy whose name is called out, blushes, but echoes the words of shame. So-and-So, *unqualified*. So-and-So, *qualified with second class*. And the boy parrots back, *So-and-So, qualified with second class*. One by one, the officers yell out; boys repeat. Whether they are man enough. Whether they are not. Their fates. Their manhood. *Shirō Shimura: qualified with superior marks,* the voice yells out. Shirō flushes, straightens his back, and he repeats loudly: *Shirō Shimura, qualified with superior marks.* He is a man. The Emperor–god has proclaimed him as a man, as a superior man. And here, right there, as he stands naked, he becomes a man. A real man in the eye of the state.

June 1936: The First Child

He didn't think he would be so happy being married until he was married. He couldn't have imagined what marriage was until he himself got married, and how wonderful it was to *live* for another, to *work* for another, and to sleep with another. In a way he never thought possible, he feels himself a better man. And when Masa started throwing up every morning a couple of months ago, when Eiko, his mother, looked shrewdly at them, herself so pregnant her belly was about to burst, and said, *There's a child,* how he rejoiced. It was a busy season last month — the beginning of the rice planting, and so much to do: to let the irrigation in to flood the land, then the hardest part: setting the young seedlings. The entire family is mobilized, the entire family, even the ones that married out, and the entire neighborhood comes out to help set the young seedlings, their backs bent, they go down in one line each, taking the young green seedling from the basket hanging from their waists, then jab it into the mud. Twist the wrist so that the seedling will root. Then they take a step forward and do the same, again and again. Masa wanted to help — after all, this was her first rice planting, and having married into the famer's family, she had to learn how to do it, though she has never been in the mud before, she has never done physical labor like this. But he said no, that she's in a vulnerable stage of pregnancy, that she should mind Eiko's baby and help around the house and prepare lunch for all the workers. And she does. That's the least she can do. How quickly she learned that all the fine silk kimonos her brother had bought for her as a dowry — *enough to last for three lifetimes* — were useless. What she needed was cotton, wool kimonos, everyday wear, and as soon as she learned this new rule, she wrote to her sister-in-law to ask for bolts of cotton and wool cloth; at night, in a dim light, she sewed the kimonos quickly. There are so many people in the house, so many that there's hardly any space or time to call her own, but she finds a strange comfort in this. She has always starved for people, for family, for love, and here there's plenty of that. Besides, there is Shirō. And their new baby.

August 1936: The Proudest Day

His nomination to the Emperor's Guard came from the mayor, then went to the city council, then to the governor and the prefectural council, and from there, to the Emperor's Guard Regimental Headquarter. And with his nomination went his background file. Everything with high marks. A Guard is chosen by family background (of the middle class or above); educational history (beyond the compulsory elementary education, and at least four years of higher education, including the Young Man's Free School, and with high marks, especially in the military training and ethics); height (must be above 165 cm); looks (of handsome and clean appearance); single and cannot be a burden to the family if serving for two years; personality (moral and ethical constitutions, without any habits of gambling, whoring, and drinking); talents worthy of noting. He passes all that — except for one: he is married with a child on the way. But that is not noted on the file because the marriage certificate hasn't been filed, and it won't, not in this neighborhood, not at this time, when the marriage becomes legal only with a child that is born alive. That is not in the file, but then, in the eyes of the state, he is not a person but a file, a record meticulously kept without his knowledge, devoid of all the things that makes a man and a woman dream at night and cry, what makes them fall in love and fall out of love. Nevertheless. He has been chosen. In a neighborhood, a boy chosen to be in the Emperor's Guard occurs every five years, sometimes only once in a decade, and he has been chosen, he has been appointed, defined by the state as the finest specimen of men, to be part of the elite force responsible to guard the Emperor–god.

October 1936: Shirō Studying for Basic Training

It is late and he has had a long day, but Shirō sits in front of his desk in his room, just like he used to when he was still in school, as he used to study late to be the top of the class. One word at a time. He mouths it. To memorize. To make these words his, just as he did in chemistry classes back in Agricultural School. And once he is done with a page, he takes up a pen and copies down one word at a time. He didn't mean to take the letter from the Regimental Office seriously: *Be sure to study these words carefully before you enter basic training.* He knew that boys from this hamlet rarely took those words seriously, and they managed well enough: two years of conscription and they came back proud with a star or two on their badges. But it was just his personality, he knows. Competitive. Earnest. And besides, he is going to be in the Emperor's Guard. He doesn't want anyone laughing at him for being a country bumpkin, he doesn't want anyone bullying him for being ... well, for being a rookie. That's what the boys had told them during their leaves: Being a rookie is hard, the second-years will find any reason to punish you, and if the training was anything like the military drills he did during Agricultural School, then it would push him to his limit, and more. His hand keeps moving, tracing one word at a time from *The Manual of Infantry Soldiers* he had bought at a bookstore in

Kashima. Or was it Mito? No matter. He keeps uttering these word as he copies them one by one. It is late already, and everyone is asleep, including Masa, eight months pregnant, sleeping so still on her back with her stomach bulging out. He knows that he will have to get up at the usual time, like anyone else, but this means nothing. He will study hard. He will be ready by December 1.

End of November 1936: The Proud Son

Shirō stands proud in front of the hamlet shrine of the Water God, the god who brings forth the water to the rice paddies, who looks after all, both the ones farming and the ones going off to be conscripts. Shirō stands in front of a large banner with his name written on it: *Shimura Shirō, Congratulations*. The entire village has come out to see him off: the mayor, his teachers from elementary school, all the famous and important men, making Shirō feel that he is important, that he is more than the village, that he is coming closer to the metropolis, closer to the Emperor — and perhaps, the entire world can be his. The villagers came at all hours of the day, bringing in money, praising him, praising his family. His father, Saburō, served everyone alcohol and the visitors ravished the food. Every man was the child of the Emperor-god, but he — only he — had been chosen to be the Emperor's Guard, to personally protect the living god, to die — yes, to lay down his life, protecting him. How happy he was when he found out that he had been chosen, the only one out of the prefecture, out of tens and hundreds of thousands of boys his age. He, the eldest son of the landlord's family, whose life would have ended here, in this small village, tilling the land. How proud his father was this past week; see how proud he is, as he stands next to Shirō, his face solemn and his head hanging in modesty. But his back is straight under his best cloth, the jacket adorned with the family crest. See how proud Shirō has made his father, how proud his mother is, how proud his grandmother is. And Masa, eight months pregnant. All of them in their best.

And the villagers, waving the Japanese flags. But they all stop, falling silent as the mayor coughs, then starts the speech. "How proud we are of Mr. Shirō Shimura, who has been chosen to be in the Emperor's Guard, the best of the best, charged with the duty to protect his majesty, our Emperor." Next, Shirō gives his speech, the one he practiced again and again in front of Masa, getting the ornate words right, taking out a word here, then adding new ones. The words fall easily from his mouth. Everyone listens rapturously. With the last word, he gives his sharpest bow, as he's seen the veterans do; the crowd cheers loudly. *Long Live the Emperor! Long Live the Emperor! Long Live the Emperor!*

November 30, 1936: Shirō Leaving Home

Masa, eight months pregnant. Pregnant as a moon, but he knows that an inanimate object suspended in the sky cannot be pregnant. He knew that this day would come, ever since he received the notice from the village office in January of this year: *This is your conscription year, be prepared. Do not go to disreputable houses, for if you get sick, you are shaming your home, your neighborhood, and your country. Do not panic and try to run away, live your life as you have always been. Do not disrupt your daily routine, but remember to keep your body, mind, and soul pure.* And in August, when he received the notice: *You have been assigned the great honor of joining the Emperor's Guards.* Masa stands on the riverbank behind their house, big. His wife of fourteen days, though they have been living as man and wife ever since January of this year; it was the way marriage took place. Marriage defined by the village, by the compatibility of two people, and the woman's ability to bear children before law defined a couple as married. Shirō looks at her standing there alone, a pace or two away from his family, still shy, still an outsider. They haven't had time to be alone. Only at night, after Shirō had been forced to drink with men, with relatives, with neighbors so proud that *their son* has been chosen to guard the Emperor–god, so proud that he is one of one or two of the prefecture chosen out of so many, going to the Imperial metropolis. Their proud son, the model man. And last night, he came home, his footing unsteady, his gait engulfed by the darkness of the night and the wintering air, and slid into the bedding they shared. She turned around, not asleep, and he whispered into her ear, making his words as light as he could, half-drunk in alcohol, half-drunk in love and earnest. *I will bring honor to the family, I will make you proud,* and she nodded along with each word, excited by his dreams and hopes because he was excited, because she was in love and couldn't imagine that her beloved could be anything else, because she didn't know what a marriage was, yet. They are still children, they still believe in each other's immortality and goodness. And the light comes too early, not enough time for these lovers, and before the cocks crow and horses begin to pound their hooves on the frozen ground, he puts his hands on her stomach and prays for a boy; she wishes for boy as well to make him proud, and they pray to the gods together to make it a boy that looks like his father. And now it is morning. She does not jump into the boat as she had done when she first came here, the way she used to lightly jump in and out of the boat like a heron; she stands on the bank, waving her hands. Shirō is in his finest. Saburō pushes the boat away from the bank with the long pole. But it is a day like no other; Shirō is the one chosen by gods to serve the country, he is the creature of the mythological past, worthy of praise, worthy of the seat of honor. The neighbors, lined up by the river banks, wish him good luck, yell how proud they are, cry out *Long Live the Emperor! Long Live the Emperor!* Shirō looks back at his house. Masa is still waving, waving her arm, waving her whole body as she holds her protruding stomach, and he waves and waves until she is a dot far away, until the boat turns left, and she disappears behind the bend.

Fig. 2.1. Shirō in 1936.

II

The Chosen One

December 1, 1936: Emperor's Guard Infantry Regiment 1,
Third Company, Second Private Shirō Shimura

Shirō walks proudly, in his best kimono, toward the simple gate of the Emperor's Guard barrack, austere, practical, artificial, the way he always imagined army barracks to look. The Guards stand in attention as Shirō and his father walk through the gate, saluting sharply in their crisp earth-colored uniforms, their red-trimmed pentagon crest surrounded by cherry blossom leaves sharply announcing that this is not an ordinary regiment, but the chosen regiment, the infantry that has the honor of protecting His Majesty the Emperor. But the Guards are not saluting him. They are saluting an officer in the Guard uniform who has come riding in on a beautiful stallion, a horse like no other, not the farm horses Shirō is familiar with. The officer does not look around, but rides smoothly, turning a sharp corner, disappearing behind the barrack. He hears his father sigh, but not the sigh he has heard so often in winters, during springs, sighs that weighed so heavy they would have sank a boat if sighs were rocks. A sigh that almost says, *Only if I were younger.* There are boys like him everywhere, all tall, all handsome, all flushed with pride. Shirō stands in line; he tells the officer his name, and another officer hands him the uniform and tells him to get out of his civilian outfit, and he does so, packing it up in the oil-skinned paper the regiment had asked him to bring in the brochure he received with the acceptance letter. He is allowed to keep photos of Masa, pregnant, and of his parents and family, and the family *hanko* because every document must be stamped for signature, for seal, for receipt, and later on, for possible identification in case he dies in battle. That is all. Everything else is taken away: his clothing, his first name, his status, what he has done so far in his life. He gives them to his father, and without a word, he walks away. He is a proud man, and he feels like he is a better self, already, garbed like the Emperor's chosen guards: hats sharper, uniforms better designed. And the change: this is only the first day, but the sergeant (*Would I look like him in a year? In two years?*) yells at the new recruits, calling them by their last names and rank, and for the new recruits, there is only one rank. Private. No first name that makes him unique, individuated, with a history of his own. He is now Private Shimura. He is no longer singular, but represents the family, his village, his prefecture. He sees older soldiers sneering when he yells out his name, *Pri-vate Shimura*, with the middle syllable going down, the last syllable twirling up. *Country bumpkin*, someone whispers. He blushes, but he is earnest and nervous, and all he wants to do is to do his best. After all, he is the best, the chosen one, chosen to do his best. He is of *trustworthy character, above middle class, good-mannered and of high intelligence, higher than six years of elementary education and four years of vocational education, has gone through 100 hours of moral and citizens' classes, has completed 250 hours of vocational training and 350 hours or more of military training at school, or is guaranteed to finish it, someone who does not have to worry about the family, and will go through the basic training with superior marks.* They have kept such a meticulous record of him for such a long time, from the moment he was born. They have been watching him, they have been following him, and now, even though he is the eldest son, even though he is married, even though he has a child on the way due any day now, even though he has nothing to worry about back home, even with all these odds against him, he has been chosen. He straightens his back as much as he can, trying to mold his body into the shape of his officers, brilliant and godlike, someone he must become, and he will become.

December 1, 1936: What the Emperor Bestows upon Private Shimura

— A model 38 rifle with the Holy Chrysanthemum insignia
— A bayonet knife
— An ammunition case
— A rucksack, sleeping bag, aluminum water canteen, aluminum cooking pot, and a tent
— A combat hat
— A set of informal uniform (for inspection, standing guard, to wear on leaves)
— A combat fatigue (for everyday use)
— A formal uniform and hat with a star surrounded by cherry blossom leaf and buds (for special inspection, for the march to go to the front)
— A rain coat and winter coat, one of each
— A white work cloth
— Two shirts, two pants, two pairs of socks, and two pairs of gloves
— A personal pouch (to be hung from the neck)
— A set of loincloths
— A pair of formal buckskin boots to be worn with the informal and formal uniforms
— A pair of cloth shoes for everyday use
— A pair of indoor shoes (leather for officers, rubber for recruits)
— A comforter (two for winter), a blanket, two sheets, a pillow, and a pillowcase
— A red badge with a gold star with cherry blossom leaves and buds

December 1, 1936: Saburō Walking Away

He wishes himself younger, to be as young as his son, he thinks, as he walks away from the gate. He goes down the hill by the Imperial Palace, all the cherry blossom trees naked because it is winter and it is colder, and spring is so far away. In his hands, he carries the package Shirō handed him, the package containing all the proof of his son's civilian past, all the proof that he no longer belongs to the family only, but that he belongs to the state, to the Emperor. And soon, as Saburō goes down the Kudan hill, he comes across the Yasukuni Shrine. He stops and nods. Why not enjoy this day, now that he is in Tokyo? Why not go to the Shrine and ask the war dead to make his son into a real soldier? And he does. He walks through the wide path, a long walkway lined with trees and monuments, along with others. His footsteps become lighter and lighter as he approaches the toride, the pride of being the father of an Emperor's Guard straightening his back, his legs moving in the rhythm of how he imagines soldiers marching. He can almost see it: his younger self, wearing the crisp beige uniform of the Emperor's Guard, adorned by the insignia reserved only for the elite force. He straightens his back. Oh how he wishes himself younger so that he can be a man, so he can be as proud, as confident....

December 1, 1936: What Shirō Has to Remember

As soon as he, along with fifteen other men, is herded into their room and shown which bed and desk belongs to whom, to drop off what few personal belongs they are permitted to hold on to — pens, glasses, bank books, seal stamp — the second-year who leads them turns around and yells, *Attention!* Shirō snaps into attention, just as he was taught to do during the military drills class back in school. And all around him, other men snap into attention, standing so straight, their legs and arms straight as a board, their backs so rigid they almost threaten to topple over. And that's when Shirō notices from the corner of his eye the drill sergeant walking in, eyeing one new recruit after another, sizing them up, staring at them. The air tenses. The drill sergeant's indoor shoes, leather, would have clicked if they were boots, but instead he stomps as he walks around the room. He opens his mouth and yells, *This is what you need to remember! And I will only say this once, so remember quickly.* Rules of how to treat uniforms; rules of how to store personal possessions; rules on where to smoke, how to wash dishes; rules on how to take bath-time: how to disrobe, how to store away discarded cloth, how to use soap, how to wash your body, how much water to use, how long you can stay in the bath, how to stack up bath stools; rules on how to conduct yourself in front of the sacred photos of the Emperor; rules on how to clean the toilets; rules on how to use toilets: how to piss, how to wash your hands after the toilet, how to wipe your ass, how much paper you can use, how long you can stay in the stall, how to tuck in your cloths, what to do in case of diarrhea; rules on when to wake up and when to sleep; rules on how to sleep...Shirō takes mental note of everything. Remember, remember, he keeps telling himself, this won't be repeated, remember. And he will, each and every detail, because he will see what happens to men when they do not remember.

December 4, 1936: The Body Is No Longer His, Time Is No Longer His

What he took for granted before he came here no longer exists: time, body, freedom to breath, to think what he wants, when to eat, what to eat, even the language he speaks is banned, and he has to learn a new set of language all over again. His father is no longer the one he left behind, but the one no one has seen — the Company Commander. His mother is the Drill Sergeant, but unlike his mother back home, this mother is fierce, unforgiving, punching his *children* for no reason, or so he thinks. The morning starts out exactly at 6 a.m., with long bugle notes that always end slightly off-key, and men next to him, in front of

IMAGINARY DEATH

Fig. 2.2. Shirō in 1936.

him, all jump out of their beds like surprised cats and make their second-year buddy's bed then their own; and change into uniforms, all within a minute. One or two have gone to bed wearing their fatigues, just so that they don't have to deal with buttons and arms. With the call of *itadakimasu,* they shovel in their breakfast — ten minutes to eat; at the end, they jump out of their chairs and yell out, *itadakimashita;* fifteen minutes to wash everyone's plates. And then from there: drills, running, drills, running, reciting, running until lunch. And if they aren't running, they are punched, kicked, for little offenses: not being able to eat in ten minutes — that warrants all sixteen of them being slapped. The first one slaps the second man in line, the second man slaps the man behind him, on and on. And if they aren't hard enough, they have to do this all over again. *Fast shitting, fast eating, fast running.* And then bath time and dinner time. All within fifteen minutes each. But there is no rest until lights out. The second-years, those dreaded second-years, come sneering in, trying to find faults, or no faults, just because. Yes, just because. Corporal punishment is a way to remold men, fists are what get men to remember the rules, rules, until they can do this in their sleep, and speak in the new language, a dialect, a completely new dialect only spoken in this new family. This is a new family, fierce, rigid, hierarchical, and the only way to survive in this family is to watch, to note, and to submit to the rules, wholeheartedly. They will think of this until 8:30 p.m., when the light is out, their sleep so immediate and heavy, while a part of them is alert, waiting to hear that one bugle note.

December 29, 1936: Shirō on His First Home Leave for the New Year

This is his first home leave, the first time he is allowed to leave the barracks, leave the confines of the prying eyes of others, the first time he can breathe what he calls *the outside* now. And how strange. It was only a month ago, but a month, inside, feels like years, centuries, each day five years' worth, as if he has aged, or perhaps as if he is living a new life, as if he had died and was born again. That's how he feels: that the man he was only a month ago is not this man in uniform. This man's back straighter, his gait more masculine, even his speech is of a different language. His home, a dot he can almost see in the distance, does not look the same. A new landscape. A new set of eyes seeing this supposed familiar sight of his origin. But he must hurry; there is a child waiting, *his* child, still unnamed, but the child born on December 23, the same day as the Crown Prince, an auspicious child, a child so fitting for the Emperor's Guard in training. How he trembled and jumped and would have sung, if he could have, when he received the telegraph: *Boy is born. Stop.* How he had to go around telling this to all the others, and, slapping his back, cheering, men celebrated with snacks because they were first-years and first-years, at least during the basic training, weren't supposed to drink, not at all. A brief moment when they were allowed to be who they truly were, not afraid of unknown mistakes, not afraid of collective punishment for one person's mistake, to briefly take on the language they were born with, not the new one, the crueler one dictated in the militaristic precision. And he still carried the glow, the lightness, that his life is just beginning, and that maybe he may be the same man that left this hamlet, now dry and expansive, the earth resting for the next spring to come. He writes later: name our son Seijurō: *a boy of purity, long life and celebration.*

January 1937: The Death of Private Life

Eyes are everywhere. Ears hear all, although which ear, he does not know. Here, in this new home, there is no private life anymore. All hours are dictated by the rules, and rules, one understood, can mold a man into a creature of habit. And he does everything to follow the rules because he is earnest, because he knows what it would mean if he does not: sometimes he loses patience with others, the ones who do not follow the rules as they should, not because of the punishments, but because he cannot understand why they don't take it seriously. It's his nature. It's his personality. He can feel himself becoming a better man more and more as he follows the rules; he can feel it, or at least that's what he tells himself. Letters arrive with black marks; letters leave with words censored out. Journals, which they are all forced to keep, are read. Soon, Shirō has learned to keep his words minimal, to borrow words of this new home. And he is efficient. He can mimic these words, he can imitate these words like a good parrot.

February 1937: A Day's Worth of Consumption

For every meal, soldiers will be given one bowl of rice, one bowl of soup, and one side dish. The government has calculated that a man needs 3,000 calories per day to be combat-ready: 900 g of rice, 210 g of meat, 600 g of vegetables, 60 g of pickled radish, 75 g of miso, 5 g of salt, 20 g of sugar, and 3 g of tea. This is all soldiers need to be battle-ready, effective, strong, and agile.

February 1937: Glory for the Family

He enters the arena, takes two stiff steps, and bows. The opponent bows at the same time. Takes three more steps, the floor cold underneath his bare feet. The last match; one more, and he wins the regimental *kendo* tournament. He grips the familiar bamboo sword and closes his eyes. The armor hangs heavy on his shoulders as if he is carrying the honor of the family. And he is. He squats down in the pose of the ritual, the way a match starts. Two men facing each other. The moment of pause with the tips of their swords crossing lightly. He remembers his former teachers' voices: keep your mind clear like still water. Do not think about winning or losing. Only when your mind is empty can you see that moment, that fraction of a second, when there is an empty space to strike. Only with the empty mind can you see that fraction of an inch on wrists or between the elbow and the torso, to strike. He keeps his mind still. The call to start. And he springs up like a cat and strikes. The roar from his brother–soldiers; the tip of the opponent's sword dropping to the floor. He has won. He is the winner of the kendo tournament: the winner of the regimental tournament.

February 1937: Dismantling of the Self

The self no longer exists. The selves that these men have nurtured through the region of their birth, through their family and their individual history, through their education and their secret thoughts are dismantled one by one through punches, through sleep deprivation, through restricted diet and rigid control of waking hours. The hours of running. The hours of reciting the Army Rescript. The hours and hours of enforced training and the constant lecturing of what it means to be a soldier in the Imperial Army. And when sleep comes, these men do not dream of home, of love and sex, but only see darkness, only to be woken up rudely with the sound of the bugle. Their language is no longer the same. Their definition of privacy — nonexistent. When an order comes, their bodies move before questioning, their gestures automatic. And they have learned to move as one, they do everything together, the self annihilated, replaced by the group. But then, little do they know that they have been doing this all their lives: with teachers, with rules, and, more importantly, in a family, the father in control of the fate of the members.

February 28, 1937: The Officer Track

He does not know this, but for the past three months, the officers have been keeping track of him: his personality, what he does during his leave, his marks, his private thoughts, what he does when officers are not around, his intelligence, his physique and sexual conduct, his morals, how he uses money. They watch him as he lies on his bed reading letters from home; they watch as he raises his rifle and aims at the target and hits all the marks; they watch and take note as he recites the Imperial Infantry Rescript without making a mistake; they watch as they make the rookies hit each other for a mistake made by one slow man, whether Shirō puts all his strength into punching that man; they read all his letters sent through the army post, and turn blind eyes to letters he has been sending through the regular post; they watch his every movement, his every gesture, and note it all down. The file he does know exists in the back office, where all the files exist: who is fit to be promoted, who is not, men categorized into soldiers and not. And now, he is the only one nominated for the officer track from this basic training unit, one out of twenty. The file will dictate his future in the army, which will last until he is forty years old as a reserve. They watch; they take note.

March 11, 1937: First letter from Shirō

Shirō writes his first official letter sent from the army post, just like he is taught. He writes as if he is imitating a model letter, just like his officers dictated, writing how gallantly he is learning to be a soldier, how wonderful the basic training is, how he misses his family. There is not supposed to be any individuated emotion, yet, he does manage by writing individually to his wife, addressing her as *My Love, Masako;* he does tell her, *wait for my return, I will be home soon,* yet these two lines are blackened. These two small pieces of information did not pass the officer's censorship. What he could not write: that he was forced to hit other men, almost as a test by the second-years, to hit as hard as he could so that no one would punch him as punishment; that there is constant emotional and physical oppression, and though he rationally understands, *if we don't follow orders, we might die in battlefields,* there's a part of him that wonders, *why must I follow stupid orders like doing one hundred pushups at three in the morning, why should I be forced to go around with full buckets in my hands, going from one room to another and yell out, One is doing this because one is stupid, please punish me;* never-ending humiliation over small things; how men say that in the army, everyone is equal, everyone can advance, but it is not true. The more stars you have, the better the chance of survival because you have more food, more space, more of everything. All of these, he sees, but cannot write.

March 1937: The Preparation for the Officer's Track

Shirō has worked so hard, doing everything the army demanded of him. He has done all that his new family demanded of him: to obey everything, even when he does not agree, from somewhere deep inside of him, a part of him that is still a civilian; he has run 20 km in full gear without complaining; he has punched men in his unit, as much as they have punched him, hard, when ordered by the second-years and their officers; his body can move on its own when the orders come, and orders come at all hours; he can shit in five minutes, he can eat in five minutes; he has washed dishes and uniforms of the second-years in the freezing weather outside, his hands chapping from the cold; he even can write in the prescribed language of the army in letters and in journals; he has exposed his private life for all to scrutinize and to judge. Now, he is in the officer's track, the only one to be chosen from the basic training unit. He hears what others say, *brown nose, goody two-shoes,* but he knows he has earned this. He will be given special privileges: a private room to study in, uncensored letter-writing privileges, a modest amount of privacy, and someone else who will wash his clothes and dishes. He is ambitious; he can see that he will go far in this place as long as he works hard, as long as he can prove himself better than anyone else. He has worked so hard, and he has earned each and every promotion, with each lacing of his boots, with each bullet he fired, with each punch he took and gave, all in the name of the family honor.

April 1937: Mr. Soldier Walking Around Kudanshita

The cherry blossoms turn the landscape pink along the Kudanshita Boulevard as Shirō walks, in full regalia, downhill. It is his day off, and he is mesmerized by how beautiful cherry blossoms are, and what they stand for: the national symbol. His saber rattles at his waist. He is in the full uniform of the Emperor's Guard, though he is not a full soldier yet. He still has three more months of advanced training, three more months of studying to be an officer in the elite corps within the army. On a day like today, when he sees families walking together, when he sees an old couple supporting each other as they slowly make their way up the hill, he wishes that Masa and little Seijurō were next to him, he wishes so much that he could show them the beauty of the sight he so loves: Kudanshita leading to the Imperial Palace, cherry blossoms, a man and woman tentatively walking close to each other, so close that they are almost touching each other, but not quite. All the things that he is meant to fight for. As he turns toward the moat circling the Palace, a small boy comes up to him and salutes, just like a little soldier, and he can't help but to smile. *Mr. Soldier, thank you for serving our country!* the little boy pipes, and he wonders if his son will eventually be like this little boy in front of him, and his heart aches for a second. So much he is sacrificing: the sight of his son taking his first step, waking up with Masa in the cold morning, being out in the field, preparing the earth for planting, the crowded and never-quiet breakfast with his parents, his grandmother, and his brothers and sisters. So much, his life interrupted to serve a greater good. On a day like today, he knows what he is and will be fighting for.

3:30 p.m., April 9, 1937: Landing in London, a World Record Is Set

Kamikaze, the plane that belongs to the *Asahi Shinbun* newspaper, sets the world record for the shortest flight between Tokyo and London, and all of London comes out to celebrate. Sixteen thousand miles in 94 hours, 17 minutes, and 56 seconds. All of Japan celebrates to congratulate the son of the nation. How proud we are, we tell ourselves, see, we *are* equal to the West. A twenty-five-year-old aviator as great as Charles Lindbergh. A navigator who is half-British and half-Japanese, the face of a white man but who speaks *our* tongue and who has our name. The plane, named after the Divine Wind that drove away Mongolian invaders twice, has been blessed by the godly wind. All of Japan celebrates the homecoming of these two sons — the aviator and navigator. Only four months later, in August, both of them, along with the plane, will be drafted into the war. They will be the property of the State. They will be one of the wheels to keep the myth, the nation, and its sacred war going.

June 10, 1937: The NCO Track

Shirō cannot write to Masa that he did not make it to the Officer Track, but instead, to the noncommission track, where the highest he can rise will be sergeant major. He thought that he could rise here, that as long as he can show that he is better, that he is the best, he could prove himself worthy to be an officer. What he doesn't know: that in order to be an officer, you need a college degree. What he doesn't know: that in order to be a career officer devoting their lives, they cannot be a first son because the first son is supposed to carry on the family name, and the family comes before the government at this time. At least for now. What he doesn't know: his father might have written to his officers asking them not to promote him because he is a first son, because he is a new father, because he has a wife, because he is needed back home. All of this Shirō does not know, and he is ashamed, he is disappointed, he feels himself not worthy enough in his wife's eyes.

July 8, 1937: Marco Polo Bridge, Start of the Second Sino-Japanese War

It only takes an unannounced firing; it only takes two sides already tense, pressure building up for the last six years, in order for the thing to explode. Two enemies on different sides of the bridge — call it the Marco Polo Bridge Incident, Lugouqiao shibian, Rōkokyō jiken — names do not matter. A small incident: a shot fired unannounced, the Kuomintang Army firing back, then a skirmish between two armies over a missing Japanese soldier. This small incident explodes at the arched bridge that Marco Polo sang about 500 years ago. This small incident ignites the fire and explodes throughout China, carried by Japanese soldiers fanning out, burning houses and firing at Chinese, both soldiers and civilians.

August 1937: A Big Hurricane

As if the gods are foreshadowing the thing to come, they bring a hurricane under their arms and throw it against the shore of Ibaraki. The rivers flood; the ocean turns gray, black; waves rise as high as the houses by the sea, swallowing houses, people, ships, and crops. The farmland in the great Kantō area is completely devastated. All the rice crops, right before the harvest, their kernels about to burst, are swallowed up by rivers that overflow the banks, flooding the floors of houses, the golden reeds drowned and washed away. A crop failure. A house by the river, the house of Shirō's birth, nearly sunk; their harvest, gone.

August 1937: Horses, Too, Must Go to War

We should have seen it coming, when the town hall began to keep records back in 1921. Now, they are drafted, just like our men, though owners are given more money for a horse than a conscripted private would earn in a year. Horses are groomed,

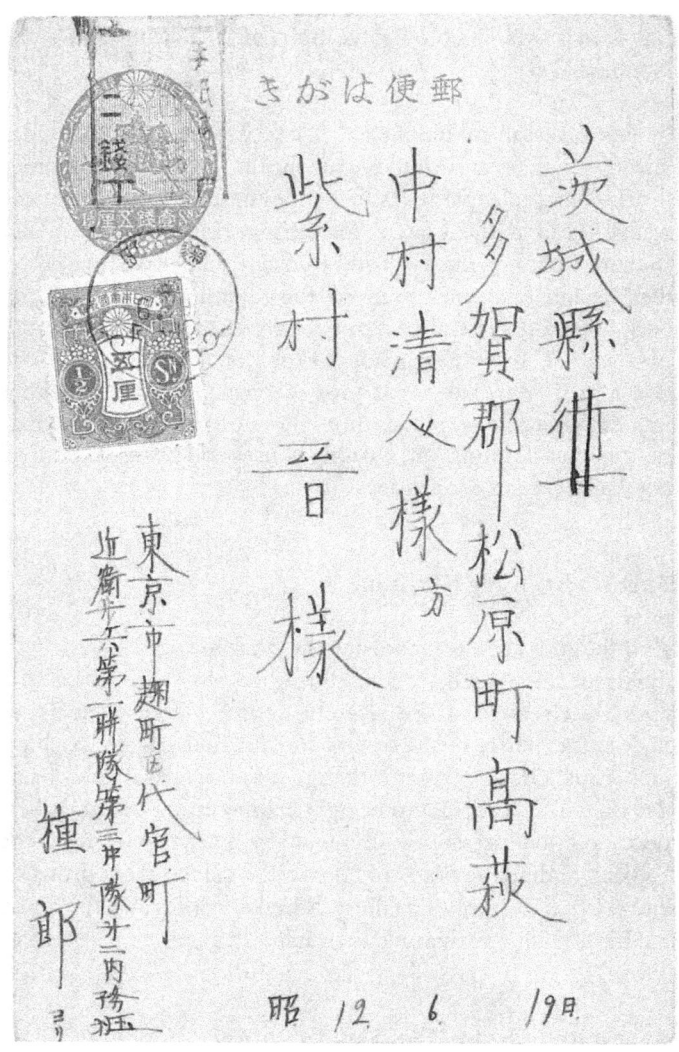

Fig. 2.3. Postcard from Shirō to Susumu, 1937 (front).

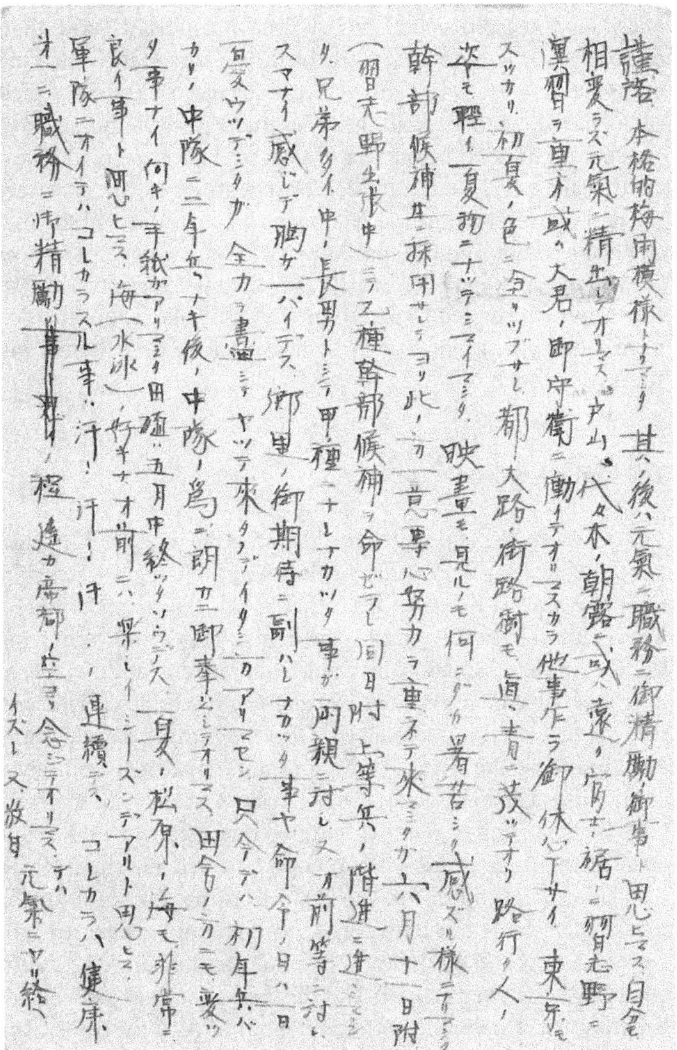
謹啓 本格的梅雨模様トナリマシタ 其ノ後ハ元氣ニ職務ニ御精勵ノ御事ト思ヒマス 自分モ
相変ラズ元氣ニ精出シテオリマス 戸山ノ代々木ノ朝露ニ或ハ遠ク富士ノ裾ニ習志野ニ
演習ニ重ネ或ハ大君ノ御守衛ニ働イテオリマスカラ他事乍ラ御休心下サイ 東京モ
スッカリ初夏ノ色ニ全ク包マレ 都ノ大路ノ街路樹モ眞青ニ茂ッテオリ 路行ク人ノ
姿モ軽イ夏物ニナッテシマイマシタ 映画モ見ルモ何ニタカ暑苦シク感ズル様ニナリマシタ
幹部候補生ニ採用サレテヨリ此ノ方一意専心努力ヲ重ネテ来マシタガ 六月十一日附
(習志野出張中)ニテ乙種幹部候補生ヲ命ゼラレ 同日附上等兵ノ階進ニ進ミマシ
タ 兄弟多イ中ノ長男トシテ甲種ニナレナカッタ事ガ両親ニ対シ又オ前等ニ対シ
スマナイ感ジデ胸ガ一パイデス 郷里ノ御期待ニ副ハレナカッタ事ヤ命令ノ日ハ一日
憂ウツデシタガ 全力ヲ盡シテヤッテ来タノデアリマシタカラ 只今デハ初年兵バ
カリノ中隊ニ二年兵ナキ後ノ中隊ノ為ニ朗カニ御奉公シテオリマス 田舎ノ方ニモ変ッ
タ事ナイ向キノ手紙ガアリマシタ 田植ハ五月中終ッタソウデス 夏ノ松原ノ海モ非常ニ
良イ事ト思ヒマス 海(水泳)好キナオ前ニハ 楽シイシーズンデアルト思ヒマス
軍隊ニオイテハコレカラスル事ハ汗! 汗! 汗! ノ連續デス コレカラハ健康
第一ニ職務ニ御精勵ノ程遠ク帝都ノ空ヨリ念ジテオリマス デハ元氣ニヤリ給ヘ
イズレ又後日

Fig. 2.4. Postcard from Shirō to Susumu, 1937 (back).

packed with carrots, carrying a little passport around their neck. The name of the horse, its birthday, personality, the kind of horse it was — plow horse, equestrian, packhorse — their ranks based on their birth, just like men. And the horses, too, are sent off amidst the fanfare of flags and cheers. For the next eight years, nearly half a million horses go off to war. They are the first ones to be shot when supplies are low; they are the first to be shot when injured and lame. The ones conscripted near the end of the war are old, just like the men who are drafted. The conscription crippled farm production, just like when the men were taken away. None of these horses came back home, but some of the men did.

The Restless Spirits

The dead who die suddenly, the dead who die without marriage, without finishing their fully given lives, their regrets stronger than the will to move to the otherworld, these spirits linger. They haunt. They stay amidst the living. They haunt the living the way mourners haunt the dead. This world is full of grief, of lives that must be lived, of hopes that need to be actualized and dreams that need to be realized. Give them three days to come back amongst the living. Give the unmarried men doll-wives so that they can be tricked into believing that they are married. Give the dead children doll-friends so that they won't take children of their age to keep them company. Wrestle them down into submission. Bind the war dead into the Yasukuni Shrine where their souls can be washed, purified, and enshrined into godhood so that they do not move, so that they do not protest, so that they will be pacified. And the living, too, will be pacified, told that these men did not die in vain but died for something bigger than themselves, something bigger than their home.

October 13, 1937: A Letter from Shirō to Masa

My dearest wife, my dearest Masako,

9 p.m. Thank you for your letters. I'm so sorry for not writing for such a long time. Foremost, I am so glad to hear that you are doing well. I know what a hard time you have gone through during my absence, with my mother giving birth, and father's anti-air-raid drills. At the same time, I am so grateful that you have been taking care of Mother so diligently. I know and understand all about your hardship, but no matter how much I wanted to, I couldn't even write a letter that may soothe your worry and aggravation because of the duties my country demands of me.... Masako, I know how you feel because our hearts are one. Know that you are the only person I share my heart with. Every day, I look at the picture of Seijurō looking so healthy. Do raise him with all your love. I believe in the days when the three of us can converse happily. I will be released soon. I don't know whether I can leave this November, as is the usual case of conscription. I don't even think about it. About three thousand soldiers have been mobilized to China from our regiment. I'm working on the mobilization paperwork. After this is done, we might be the only ones left in Tokyo. Be prepared that I may be sent to the front with the next mobilization. I may be sent tomorrow. For the big autumn training, we might join the battlefront in Mongolia or Nanking. I've already given my body to the Emperor; the only thing I can do is to wait for the order. I am well prepared to go. Do not worry about me.

Masako, you take care of yourself. I hope you do not do things that will make people point fingers at you and laugh. I have told my sisters about the possibility of mobilization already. I know you have been ready for this, but I wanted to drop a line to let you know. I will write a letter to Big Sister Haruko as well. I also saw Gorō for three days and he is doing very well, so don't worry about him. Grandmother is old and she might worry needlessly, so do explain to her and tell her not to worry. Busy harvest is over, and so are the anti-air raid drills and Mother has already given birth, so you now rest and get your strength back.

A quickly written letter to home,
Shirō

P.S. I am now smoking about five cigarettes a day. I wanted to let you know — my dearest Masako — about this.

October 12, 1937 — November 11, 1937: Shirō Goes to China

His first trip abroad, ever: to China. On top of the ten-*yen*-five-*sen*-a-month salary for a Private First Class, three *yen* is added for "battle salary," making it fourteen *yen* a month. A month. The month when the Total Spiritual Mobilization is decreed by the government, when all citizens must put all their hearts and souls into the war effort. For Shirō and his buddies, this is the end of the first year. All the stories he heard from the second-years, from his officers, of battles. Of fights. All the men they've killed, as if killing made them better men, and it did. In the army, killing does make men into real soldiers, and in their eyes, the soldiers who have not killed are not worthy to be called soldiers. But it is a gentler war still, where the Japanese army is only fighting against the Nationalists, and leaving the Chinese peasants alone. That's when they still tell their soldiers, *this is a holy war, a sacred war, to free China from the backward warlords; Asia as one, in peace and harmony.* Shirō and his unit are thrown into the front, but always positioned away from the front where they can die, where a year's worth of training is wasted if these men were to die. So they are positioned where bullets can be heard, but not close enough to hit. And how scared Shirō is, how scared all his buddies are, as the second-years laugh at them; how they cower in the foxholes like little boys as mortar shells hit all around them; how they hear injured and dying men screaming for help in other trenches, but they are too scared to do anything, the only thing they can do is to hold onto their bowels but even these betray them. He learns that artillery and mortar shells are deafening, a body blown up into pieces, bone fragments and bloody globs flying through the air, almost becoming ammunition to kill others; he's even seen men knocked unconscious by someone's head, forearm, foot, striking them. He learns that bullets are clean, making perfect red circular holes, but they are messy, that it's rare to be shot in the heart or in the head; most men don't die, not immediately, not for now. He also learns that bullets have their own music; Chinese bullets sound this way, Japanese bullets sound that way, the distance can be measured by their music. Why, look, some of the battle-hardened soldiers can even sleep through battles. This is the month when Shirō does not write a letter. He cannot because he does not have the time, because he does not have the words, because there is no word for the deaths, the carelessness, and the brutality he sees all around him. Three *yen* covers the experience in battlefield. For all the horrors, three *yen*. Yet, somehow he proves himself in the chaos of bullets, in the time of blood and dying. At the end of the month, he is promoted to the rank of a corporal; he will earn 13.5 *yen,* and he enters his second year.

December 13, 1937: The Fall of Nanjing

The war in China was meant to last only three months, but it's been nearly six months since the shot rang across the Marco Polo Bridge. The war in China hasn't gone all that well. Everywhere the Imperial Army goes, they can't tell who are friendly, who are spies, and even children can be working for the Nationalists, even women. Marching days on end, carrying all their gear, all twenty kilograms, because trucks are rare and only officers can ride horses. They may steal a donkey here or there, and sometimes force Chinese prisoners of war to carry their gear, but no matter what, no matter how many battles you've fought in, what matters at the end is how many stars and stripes are on your uniform, not how many years. The young officers on their horses, so few battles under their belts, get tents, better food, more food prepared by cooks, more comfort, while the infantry walk, walk, and are told to find their own food. Supply lines are thin; they haven't caught up. Even letters haven't caught up with them. And that's what they do after the day's walking: go into Chinese peasants' houses, ask for food. They do not see that Chinese peasants give up food because they are scared. They do not see that other units have already gone before them, and that there is nothing left in the rice bin, no ducks, nothing. The Japanese Imperial soldiers are told to walk as quickly as they can, "Six Kilometer March," six kilometers per hour, walking eight hours a day, nearly fifty kilometers a day, in full combat gear. Two men are said to compete for how many heads they can decapitate, and of course, they exaggerate when they are written up by newspapers. They are scared, they are hungry, they are exhausted. And there, these civilians who could be Nationalists, Communists, who glare at them, who are defiant. By the time they reach the outskirts of Nanjing, all they want to do is to let go, to do harm, to punish the *Chinks* who killed their brother-soldiers, to show them who is the master, because they themselves are no longer masters of their own lives.

December 1937: Shirō Training Men for the First Time

Instead of going into battle, and fighting in the war, the army tells him there is another way of serving. Instead of being sent to be stationed to different posts, to serve the Emperor around the Palace the way the Emperor's Guards are meant to, he is told that he is to train new recruits under the drill sergeants. His regimental commander — a father, almost, who never talked to corporals and sergeants, like the patriarch in the family — had lectured to selected NCOs chosen to be drill sergeants, that only a handful is chosen every year, that it is a great honor to be chosen, that they are the ones deemed extraordinary enough in conduct, in morals, in skills, in intelligence, to teach the generations of rookies. Shirō feels honored; even if he wasn't good enough to be an officer, he is good enough to be the next best thing: to be a drill sergeant. And as he looks around the fresh faces of boys, exactly the way he was a year ago, coming to the basic training barrack with his father, remembering how he was in awe of an officer riding out on a beautiful horse, here he is, welcoming the new recruits. They look so young, he thinks, so young and scared, still like a *civilian,* they even walk like one, he thinks, and smiles to himself. It is now his job to make these boys into soldiers. It is now his job to do everything to dismantle the individual in them, and remold them into the soldier, as he is now. But there is another part of him that wonders whether he is really a soldier yet, because he has not yet fought in a battle, because he has not yet killed. But for now, he is on a noncommissioned officer track; his job is to hassle these boys, to break them down with fear, make them feel so powerless that the officers will be able to build these boys into Imperial soldiers.

January 1938: After the Fall of Nanjing

They say that one prostitute will be needed for every forty soldiers, but if there are no prostitutes, then men will start raping. And then what would the rest of the world say? Soldiers have fought well, but forty rationed cigarettes stamped with the Emperor's chrysanthemum insignia aren't enough. Remember the Russo-Japanese War, when nearly half the men were rendered useless by venereal diseases when they were set loose amidst the civilians. These *Chink sluts* must have venereal diseases; they all do. And remember, it only takes one polluted slut to destroy the entire battalion. We can't let that happen. The eyes of the other nations are on us. What would they think if they see that the god's Imperial Army is raping Chinese women left and right? They would lose confidence in us immediately. They would say, *These Japs are no better than the other Orientals. That's why they need us to guide them to civility.* Give them women, that's what all men need anyways. Give them clean women, at least. Let them see that we are treating them well, that they are well taken care of: we provide women, we provide a good postal service, we provide comfort stations, we provide water. We are, after all, fighting the sacred war, and we have the god on our side. We know what men want. They want women. They all want women after a good kill.

February 1938: The Sacred War

We do not know when it started, but somehow, somewhere, the China Incident is revered as sacred, as a divine war fought against the evil of peace. It is the Chinese, the newspapers say, they are so like children. How we tried to help them, because they are young, because they don't know any better, because they have been lied to, too uneducated to see what they've done to their own country, and how they've paid us back by spitting on us. Now the Emperor is a God, the war fought in His name is the Sacred War, the soldiers, the Imperial soldiers, the children of the Heavenly Son. We are in a divine war, a decisive war that can bring peace in Asia for the next hundred, thousand years, or China under the West, China under the Communists, the unrest and the next millennium of war and devastation and servitude.

1938: What the Soldiers Leave Behind

A last will to make sure that you do not burden others with your lack of organization. Imagine how awful it would be if every time some questions came up, they have to contact you. You wouldn't be able to concentrate on your duty. Know that once you enter the service of the Emperor, your loyalty lies with him, not with your family.

A strand of hair and a fragment of nail; do not think that your body will be sent home. If you are a real warrior, you do not think about what will happen to your body, or your life, on a battlefield. Your family will already have a part of your body to bury.

A photo portrait of yourself so that your family has a photo to adorn the altar.

A military notebook that details your rank, unit, height, weight, to be left with the home regiment.

Do not worry about the rest. If and when you die in action, the Emperor will look after your family.

1938: Soldiers in the Sacred War

The Emperor's children march toward the battlefields in China. The Emperor, the merciful father and god, tells His children, *do not worry, you are taken care of, you and your family.* Of course, He does not speak directly; He has prophets to do that, like any god. First, so that soldiers can stay in touch with families and friends back home, He sets up a postal service and gives each and every man a handful of postcards and envelopes with pictures of stars and pigeons with helmets on. *Write,* the Emperor commands, *write so that the family does not worry about you, write so that they can know how much honor you are bringing to the family and to the village, but do not write about where you are, how many men have died, or your personal thoughts of the war.* Letters and care packages from home are forwarded to wherever they are, the only thing you need to do is address the letters and packages with the name, the regiment code, and company name. That's all you need, and the packages will arrive safely to the men. If a soldier is orphaned, the Emperor–god will give you the ones sent by the Great Japanese Women's Committee or the ones sent by schoolchildren. The Emperor–god is all-knowing. He — or at least His prophets — knows that men cannot survive without women, so they set up comfort stations in towns and villages considered green zones. The prophets do not like thinking of bodies and flesh, so they ask the private citizens to carry out their orders, and they do. Women are cheap, especially in enemy land. Especially when there's a colony. With that taken care of, the prophets make sure the families back home are taken care of. In case men die, and men do die, even in the holy war — as long as they die gallantly, full of sacrifice, they never die in vain — the Emperor–god will take care of their families for the rest of their lives, as long as the widows do not remarry. They make sure that soldiers have sanitized water; what is a soldier when rendered useless from diarrhea but an extra mouth to feed. They make sure soldiers are protected by giving them *Attack Number One*, condoms to sack themselves in or sack their private items to keep them waterproof. In five years or so, when the war worsens and men are scattered everywhere in China and Asia, when things become tightly stretched that holes begin to appear in the fabric of the Empire, the prophets give the order to soldiers to *provide their own*, a sacred permission to pillage and loot. The Empire is proud: see how the children of the Emperor are taken care of, their family safe in the hands of the god. The Empire closes its eye to the men, exhausted, hungry, slowly turning into hardened machines, and when things become worse, it leaves them to die. The Emperor closes himself behind the sacred veil. He is a god, after all. Let the god sleep.

March 1938: Shirō upon Hearing That Others Are Going to China, but He Is Not

Men he has trained with, his brother-soldiers, are mobilized to the Chinese theater. He is not. And he does not know what to think. A part of him, the soldier part of him that has been a part of him for the past fifteen months, is disappointed, is sad that he can't live the adventure he has heard from his older-brother-soldiers, the third-years and officers who boasted about their escapades in battles, the vast Chinese landscape, how easy it is to kill, *oh, I killed four or five, I've lost count; pow, and they fall like dolls; by the time the year's over, we won't have any Chinks left to kill.* Of course, he does not know that the third-years and officers who are here in the barracks haven't really seen war yet, not the way they boast about it, not the way they laugh and exaggerate about heads exploding, about body parts strewn all over the battlefield because they haven't gone to fight, they've been left behind, just like Shirō will be. But there's a civilian part of him, the part that is the husband, the father, the eldest son, the farmer, that does not want to go, that does not want to leave behind his family, that worries about the farm and everything that it means. Two sides of him are at war, the soldier and a man. A war fought in the internal terrain, and no one side is winning or losing yet.

March 28, 1938: A Letter from Shirō to Masa

My dearest Masako,

Spring is here, finally. The majority of the cherry blossoms in Tokyo will probably open their lovely buds any day now, maybe even tomorrow. I'm sorry I haven't written in a while. I'm glad to hear that Seijurō is doing well and that he has begun to walk about. I imagine that you, too, are doing well, helping out with farm work as well as housework. I'm, as always, doing well. The new ones who completed the basic training haven't received their orders to be shipped out yet. I heard that there will be mass mobilization of noncommissioned officers at the beginning of April, so I am preparing myself, in my mind, for that. It's strange to think that a conscript like me still hasn't been sent out to the front yet. I'm sure if I had volunteered, they would have sent me immediately to the front, but I'm still debating about it. In order to pay back the kindness of the company commander, I debated about volunteering, but once I go, I'll have to work abroad for two or three years. I thought about options, but I've decided to stay and wait until a new command comes. My life is up to the winds of history, but I'm a soldier. Do be prepared for anything. This is my only wish to my beloved wife. In the next mobilization, some of my colleagues are getting shipped out (conscripts and volunteers). Half of the noncommissioned officers will be going. Do not tell this to anyone. I shouldn't be left behind here in Japan just because I have you — my dearest wife; I cannot do that. I should not be the obstacle to myself. I feel bad that I have left you alone, by yourself. However, as a wife of the soldier of the Imperial Army, do be prepared for everything. At the same time, I don't think I will be going. There's good days and bad days. Though now we are so many miles away from each other, do believe that like spring arriving every year after winter, we will be able to enjoy our lives together. I thought of volunteering, and making you a real soldier's wife, but my conscience will not let me. I dream of returning to my home, Kashima, and serving my parents and the family and working toward bettering the farm villages and wrestling with nature and being with my family. I think that men are better off living in the countryside with nature as their enemy, rather than living in the gaudy technological capital. What do you think, my beloved wife? Every time I ramble off like this about stupid stuff, I think how self-centered I am. Do not worry because I don't worry about you. My mind is made up; I'm ready. Compared to men who left this capital overflowing with flowers and are now battling in the frontline, I feel guilty writing this letter.

I've rambled on and on.
Shirō

P.S. How are you? Are you doing well? Please look after Seijurō's well-being, too.

1938: Soldiers on the Warfront Worrying about Home

Soldiers on the warfront keep hearing the worst: so many restrictions at home, not enough matches, not enough food, not enough coal because soldiers are using up things that are needed at home as well. Women back home sacrificing their time and coupons so that they can send care packages to the front. And letters do not say anything except that all is well at home. Suspicion brews. But are they also not writing letters home that all is well? *We are doing well, there's not much fighting going on; life is the same as it was back home.* Lies keep them apart. But these are white lies, lies to keep going, lies that do not harm anyone. But on their leaves back to the *innerland,* they see that the letters from home were right. Tokyo is the same, as if there is no war going on. Women walk on streets, arm in arm, in their best fineries, laughing, joking, and giving disgusted glances at soldiers looking ragged and bewildered away from battles. In China, their brother-soldiers are dying; in Japan, the world is still bloodless, these women walk with their hands clean, and no one is hungry. Women and men dance to the Tokyo Ondo, men stand on street corners talking of politics, though they hush their voices when they speak of Communism; cars drive around without a care for petrol. The imperial city is booming, dancing, drowning in money. Soldiers see that while they marched night after day after night nonstop in the mud, in the cold, attacked by bluebottle flies and mosquitoes, toenails falling off like rotten tomatoes, scared that every Chinese person, old and young, is their enemy, here in Tokyo, people are enjoying all that the victories in China brought, not caring, not understanding, that the war is fought one inch at a time.

April 1938: The Report Compiled by an Army Doctor

… as soon as I landed on the continent, I was struck by how Japanese soldiers could not control their sexual desires, and during my year here, I could not shake off this impression. However, strangely, I never heard of headquarters prohibiting the men from indulging in this vice; moreover, they constructed comfort stations run by the army, providing prostitutes to our men. Venereal diseases spread from these women to the soldiers.… Not only low-ranked men but officers suffered from these venereal diseases and were treated in secret. Instead of contracting diseases from the Chinese population, they constructed comfort stations and staffed them with Japanese as well as Korean prostitutes, but ironically, they were the ones who spread the diseases. The army knew that they could not suppress the sexual urges of our men, so in order to prevent the raping of Chinese women, they constructed comfort stations, but the raping did not stop. Good Chinese citizens cowered in fear whenever they saw Japanese soldiers. Officers led their men to comfort stations and encouraged them to visit, while men with conscience who knew what was being carried out at the comfort stations secretly looked down on the army. There were some officers who punished their men for not going to comfort stations, calling them crazy. According to them, because battlefields are desolate, they make men aggressive, and in order to suppress their aggression, the army provided the best solution: allowing the men to sleep with women.[1]

Manuals: How to Live Your Life

There is a manual for everything: if you are a woman, there are manuals on how to write letters to soldiers abroad, both familiar and unknown; how to pack care packages so that soldiers aren't disappointed when they open the packages; how to be a widow; how to sew anti-air raid headgears; how to give birth in bomb shelters; how to turn a small plot of garden into a vegetable garden. If you are a soldier, there are manuals that teach you how to give farewell speeches on the day you leave for basic training; you will get a list of books, *Soldier's Field Manual, Rescript for Soldiers and Sailors,* to study meticulously and bring with you to the basic training; how to write condolence letters to families of the war dead; how to write letters home in accordance with the military regulations. Everywhere, articles praise the perfect war widows who do not cry but rejoice at their husbands' deaths, their faces stoic and their conduct meticulous; the articles praise the women who remain widows, raising their children on their own to follow their husbands' footsteps. Children are

---

1 "Hayao Torao Gun-I Chui Houkokusho 'Senjo Shiinkeisho Sai ni Hanzai ni Tsuite"「早尾虎雄軍医中尉報告書「戦場神経症竝に犯罪に就て」 [Army Doctor Lieutenant Torao Hayao's Report on Criminality Caused by War Fatigue], http://d.hatena.ne.jp/gurugurian/19380401.

shown how to be the perfect little patriots through textbooks, through teachers' words and punishments. And we are all faithful because everyone else is, because we know that if we show dissent, we will be ostracized, food rations cut off, the neighborly assistant cut off, and to keep quiet because we can't trust others. We close our eyes, we lie on our backs in the dead man's float, and let the current of history carry us downstream, though we do not know where we are going.

1938: A Man's Worth

A man is worth nine *yen* a month, or the equivalent of thirty kilograms of rice. No one eats that much rice in a month. Not back in 1938. That would have lasted a year back then. But if you are a man, and if you have been chosen as the child of the Emperor in the Imperial Army, you would start out at that rate. And if you need a good fuck, because you're tired and sick to death of fighting one battle after another against an invisible army, day after day, scared shitless but you don't say it; day after day, slicing off the fingers of your brother-soldiers to carry them around, so that later, when there is no battle, you can cremate them all to send back the ashes home; when the mortars explode all around you and you pray to gods, any gods, because you are shitless and there's nothing you can do but wait to die; when you have lost your way and you don't know what is real and what is not real anymore, and you need something, someone, you can buy a woman — a comfort woman — for one-and-half *yen*. Thirty minutes. You can buy a cheaper one — *a Chink whore* — for one *yen* for thirty minutes. With your salary, four women a month. But no one can live off women only. And if you are the breadwinner of the family, your savings books mean a lot more than women. A cheap fuck. Remember: you still need to buy extra socks, extra shirts, alcohol, and cigarettes. You can write home and ask for these things, if they are well-off, but more than likely, they aren't. In this war, whether you are an Imperial soldier or woman serving as the *Imperial girls' force,* your life does not amount to much.

August 1, 1938: Shirō Promoted to Sergeant

He is still under the vigilant scrutiny of the State and his officers. But he has proven himself again and again: the promotion to Sergeant rank, one of the fastest in the army. Four ranks in eighteen months. He does not know that he is a good assistant drill sergeant because he is earnest, that he is imitating the way his own assistant drill sergeant treated his group, with sincere belief that these men must be taught how to be a soldier with fists, because words are not sufficient. But he does not raise his own hand, that's what he learned. He has them punish each other, to learn that one man's mistake is everyone's mistake, and that they need to be each other's keeper. And when he does need to punch someone, he tells himself, *I'm doing this to make a better man out of him; if you can't do the basic movement of a soldier, what's going to happen to you when you are in a battle.* Of course, he doesn't admit to himself that he himself hasn't been in a real battle either. That all he is training — bayonet thrusting, hand-to-hand combat — all that may have no meaning where these boys are going, where artillery and mortar shells, machine guns, all of that dominates the battle, not men with rifles that are unstable, handguns that need close range. But he pushes these thoughts aside. His job: to make soldiers out of these men. He reads their letters and journals, writing in files about who they are, what they are, getting inside their minds and history to make the best soldiers out of them. And, as an NCO, he has privileges: a private office, letters sent without being censored, more leave days, better food, a retainer to take care of his needs, more pay. All of that, and he learns that the higher the rank, the better the chance of survival, even at a time when there is no fighting, no war.

1939: Shirō Called to the Front

Another rumor: he will be deployed in March. Maybe May. He so wants to go, now that he is the only one left amongst his unit. Top of his class in army ethics and morals, top of the class in rifle target shooting, top of the class in kendo, and a true believer in the Empire. (But not good enough to go to officer's track, he thinks bitterly, and pushes this thought aside as quickly as it comes.) On lazy nights, when they don't want to hassle any of the rookies, when they leave it to other second-years to play with the scared boys, NCOs and officers sit around drinking, smoking, maybe listening to music, maybe playing cards, and they complain about being stuck here, where they have to clean the mess the rookies make, day in, day out, one generation of rookies after another, but no action. And he remembers one NCO saying, after too many drinks, after all the bullshits and bravados are gone and the naked man sits, *I gotta family waiting for me, I just want to stay alive so that my kids'll have a daddy around,* and everyone got quiet, and looked away from each other, each in their own secret thoughts that moor close to their hearts, and Shirō, as he took a deep breath, felt a pang, a moment of realization, that that's what he, too, wanted. Still wants. And someone laughed, and they all put on the masks of masculinity, and swapped stories about this or that stupid rookie.

An Old Japanese Saying

Give us three years of failed harvest but do not give up three days of war, an old man mutters. Give us three years of hunger but do not take our men away. When a man is torn from the land, there are not enough men to toil the land, the gods are angered at being abandoned. So like children, and in their bounty and in their anger they are children, simple in their punishment, loving in their bounty, as long as they are loved. Give us three years of failed harvest and we will sell the girls for the price enough to survive the season; we will sell them so that they can shoulder the debts and so we can live a little longer. But give us three days of war and we are done for.

April 1939: Shirō Deployed to the Front

In a month, he will be sent to the front. He has been reassigned to the 212th Regimental Unit Ninth Company. He was one of the NCOs to attend the flag ceremony at the Imperial Palace, when the Emperor handed the regimental flag to the Regimental Commander. To China. When the order came, an image of Masa and Seijurō flashed in front of him, and he pushed that aside quickly. He saluted and thanked the commander but he could feel his heart beating fast, as if he was harboring a scared sparrow in his rib cage. He tells himself, *I am a soldier, I knew for a very long time that this would come;* he tells himself, *my family is taken care of, and even if I die, the government will look after my family, they will be well looked after. I am a soldier, I am an NCO, my job is to fight for Japan, and to die for Japan. As a Japanese man, what better way to live and die than to live and die for the country.* And in other moments, he sees brother-soldiers he saw in China, in that brief month that lasted for such a long time, screaming as they lay on yellow earth, and the company monk following the lines and lines of soldiers with a donkey laden with dismembered arms, all with names; to keep track of which arm belonged to whom, brother-soldiers carrying their dead friends across the battlefield so that the bodies at least can go home, if not with souls; brother-soldiers cutting the fingers off of the dead because they are so far away from the base, because they are on the march, because there is nothing they can do but to put fingers in a tobacco can; the march of soldiers clad in white bearing the boxes with urns in them, one live soldier for each of the dead; soldiers in hospitals when Shirō accompanied this or that royal family member, a room filled with men in various stages of dismantlement, an arm missing, legs missing, their limbs abruptly cut off in the middle, like photographs torn. *If I am to die, don't let me be like them,* he had thought as he stood impassive, his face expressionless by the hospital door while this Princess or that Princess went from one bed to another, *thank you so much for your sacrifice.* Will he come back alive? Can he come back alive? He does not know.

April 28, 1939: A Letter from Shirō to Masa

My dearest wife, my dearest Masako,

Thank you for the nice dinner when I was on leave. I do not think we had a chance to talk as much as we wanted when we were together, because there were so many conflicting emotions: happiness and sadness. Writing this makes it sound as if I am talking about someone else, but that is really not the case. We both know that. For a long time, I kept telling you that I am going, going, but finally I am going to the front. Now that I am leaving, I am ready to serve the country. From now on, we will not be under the same Japanese sky; it won't be the same as when I was in Tokyo.

Masako, you are the wife of a soldier. Stand firmly in the homeland and take care of home. I don't think Grandmother and Father and Mother will do mean things to you. I've taken leaves every time I had the chance, and I know your relationship with my parents is the least I have to worry about. I hope that you are serving them as I would for both of us. I think it will be wrong to say that we are unfortunate because we are separated. Do not forget that good things come after hardships. See the good, the joy, in hardship; that is the highest form of happiness.

I am planning to pull myself together and focus on fighting, and I will think of you and Seijurō when I get tired and I'm struggling. Even if we are apart, as long as our hearts are one, I know that we are more fortunate than married couples who are together but constantly fight.

Goodnight.
Shirō

May 6, 1939: Shirō Going to China

He has sent many photos home of himself in his uniform, but this time, he takes a photo in the studio, a formal sitting, the way he wants to be remembered by Masa, by his family. An Emperor's Guard. A picture of him to adorn the altar if he does not come home. He also clips his nails, just like he has been told by others, just in case his body does not make it home, and some bodies never come home. He has written his last will, because on the battlefield, there is no time to write the last words as men lay dying. All of these are left at the Regimental Headquarter to be sent home in case he does not. He is ready to go to China. He has been ready. He is ready to fight the sacred war, to fight this holy mission, and to die, if he must.

Fig. 3.1. Shiro in 1939.

III

The Sacred War

May 7, 1939: Leaving for War

He still sees her in his mind's eye as he closes his eyes and pretends to be asleep on the bottom tier of the three-tiered bunk beds on the boat, *Denmark,* as they moved away from Shibaura Harbor, along the coast of western Japan, keeping the land to the left to circle around at the southern tip to the Sea of Japan, then northward to China. Shirō and his regimental unit left Sakura two days ago amidst the fanfare where Masa and he said goodbye, took the train to Shibaura, staying at houses that opened their doors and rooms for soldiers to stay in, then marched in their finest uniforms from the barrack in Shibaura to the harbor down the boulevards where thousands and thousands thronged, cheering them off to war, *Long Live the Emperor! Long Live Imperial Japan!,* brother-soldiers marching down the boulevards just like they had trained to do, their backs straight, their steps in sync, kicking the earth as one, their rifles strapped on their backpacks. And the harbor. He still sees her in his mind's eye as he stands against the railing on the dock along with the other men, all craning their heads to see their loved ones, one last time, though they all know it is an impossible feat, women and men the size of ants down on the harbor. But Shirō still tries.

He scans through the multitude of faces, all looking up, all with smiles, amidst the fanfare, amidst the drum and brass marching sound gallantly sending men off with the beats of masculine precision, scanning quickly over the faces of the Prince and Princess, he scans and scans from one face to another, settling on a face for only a fraction of a second, moving from one face to another, for *her* face. Faces hide behind frantically waving flags, thousands and thousands of the flags of the Rising Sun, the red sun against the white. Where are they, he feels irritation rising, an irritation toward himself for not finding them, irritation that they said they would come but making it hard for him to find them, and he searches from one face to another, quickly, as the fog horn rings through the clear sky, once, twice, to announce the departure of the soldiers, the weight of the patriotism and their possible nonreturning that no one talks about. And suddenly, his eye falls on a woman and an old man with a boy on his shoulders, and Shirō stops. He focuses. There they are: his Masa looking up to the ship, his father looking up, and Seijurō looking up perched atop his grandfather's shoulders, wearing the cloth Shirō had sent from Tokyo, in the pair of leather shoes, Seijurō craning his neck left and right as if he would be able to see what the old man cannot. Waving flags along with the others, frantically, in rhythm with the others. He leans over the rail and waves his arm, waves, trying to catch their eye, so that they would see him at this finest hour, so that all he has sacrificed will be revised at this instant in their minds, and in turn, his, just because of this moment. But they do not see him. He keeps his eyes on their faces that keep appearing, disappearing, amidst the flags. The horn announces their departure. The anchor is pulled up with a steely groan, and people down in the harbor break out in unison, *Long Live the Emperor! Long Live Imperial Japan! Long Live the Emperor!,* the flags waving, hands waving, faces disappearing, and Shirō keeps waving with all his might, just to let them know that he is here, that he will come back alive, and that if this is the last time they see him, so be it, so gallant and heroic he is. And he is abruptly woken up by a call, *We are docking in Ogura, get ready, men.* He is back on the ship on his way to China.

May 13–17, 1939: Qingdao and Jining

While in the motherland they were treated like kings, heroes, welcomed with fanfare of flags and people lined up on streets, feasts readied for them everywhere they went, with beautiful young women in aprons serving them tea, calling them *Mr. Soldier,* treated like gods because they are, after all, the chosen children of the Emperor, the ones who will liberate Asia from the Western imperialists, the ones who will free China from its corrupt government. The travel to Qingdao took six days in the stifling cargo hold, three tiers of beds from the floor to the ceiling, horses moaning as men groaned from the heat, from boredom, from lack of privacy and seasickness. They stopped twice, once in Ogura to get coals; another time in Toyota and spent the night sleeping on the floor of the silk factory. They sang, they drank, they puked and shat on the floor; they tried to move their stiffening joints that creaked; they walked unsteadily around the ship when they were released from their hold for a few hours. And after six days, they arrived to Qingdao, the former German colony and now the green zone for the Japanese Imperial Army, the beautiful city of red brick amidst a European landscape. Straight roads cut out of stones that ignore the natural terrain; colonial buildings occupied by the wealthy and the Japanese; a city that overlooks the green-blue bay. And beer. Beer inherited from the Germans. How they drank beer, the thick amber-colored jewel of a liquid, as they sat watching small Chinese women walking to and fro on the busy streets. And Shirō is amazed by another face of China, the European faces mixed in with the Asian features, a woman with eyes like Masa but with amber-colored hair and pale skin like the underbelly of fish. A dead fish, he wonders. But this is the city where there is more of Europe than China, streets wider, cleaner, even the air smells sweeter, not oily, not like the way his China was last year. How he wishes he can show this to Masa and Seijurō, but letters will have to wait. Four days later, they are told to assemble, to ride on the cattle train, *we're entering the red zone, arm yourselves.* He clutches his rifle as he sits on the hard floor of the train as rain falls on him, the distance between his brother-soldiers and him so close that they can smell each other's breath — unwashed and sickly sweet — yet he does not know these men, not just yet. He has been taken away from the squad he has been training with, thrust into a newly formed regiment. He cannot trust anyone, not just yet. He sleeps with his arm under his head as a pillow, his hand holding on to the loaded handgun, just in case. Just in case the train makes a sudden halt. Only a night of this. And they arrive in Jining, the dusty dry fortress town west of Qingdao, a fortress town surrounded by four high red walls and a gate, still considered a green zone, though precarious in the balance of power, the allegiance of the Chinese as precarious as the zone itself. Who is the enemy. Who is not. They are told never to trust the Chinese here, now that they are in the battle zone.

Mid-May, 1939: A Letter from Shirō to Masa

My beloved Masako,

It was very nice to see you while I was on leave. Nothing makes me happier than seeing you doing well. Seijurō has grown so much, as well. I am very busy right now, so I'm sorry I haven't had a chance to write to you. We arrived at the new guarding post yesterday. We left from Sakura and stayed in Kobe for a week. We had about 50 men, so it was difficult to house all of them. We left Kobe harbor, where it was filled with so many people seeing us off. A quiet sea voyage and we arrived at —— (different from the previous one), and finally arrived here yesterday. I was relieved to have done this serious duty, but once again, it's the new recruits. This is their last training, so I will do my best; don't worry about me.

The package arrived safely, so do not worry. Since there's a bolt of cloth, I can make a vest or a bag. Do thank Big Sister for me. The Village Committee sent me the family photo. It looks very nice. Have you sent the packages I asked you to send to my instructor in Kanda and the company leader in Shibuya? I'm sorry I'm asking you to do so much. Can you also send me two pairs of hollow wooden clogs? I received a bonus — 20 *yen* — as well as 5 *yen* for Seijurō. When I have time, I will send it to you. I'm so busy I don't have time to write more, so this is it. My adorable Masako, it's getting cold so do take care of yourself for my sake as well as yours. Take care of Seijurō, also.

From Shirō

1939: What the Soldiers Want to Believe

You become a god when you die in war. No one will forget you — after all, people may forget the names but not the gods themselves. Your family will be honored as *Homare no Ie* — the Revered Home. They will receive a letter from your commanding officer, praising how gallantly you fought until your last breath or how you threw yourself in front of your brother-soldiers to save them. Don't worry, you will never die a miserable death in this war, in this army. You are a chosen child of the Emperor–god; you are invincible. Your cause, a just and heroic cause. If you leave a wife behind, the Emperor–god will take care of her. She will be compensated, first with consolation money, then with a pension, all based on your rank (do take care to rise in rank, because your family, too, will be affected). Your body will be retrieved. Be sure about that. No one will leave a body behind to be buried in the heathen land. You will get a proper salutation once your ashes return to the motherland. Twice a year, the Emperor will come visit you and pay respect at your new home at the Yasukuni Shrine; your family, too, will be given a round-trip ticket to come see you. But here's a trade-off. No one in your family will be able to mourn for you in the open; none of them will be able to put you under the family gravestone. You will no longer have your name. You will be buried in the corner of the plot, under a pointed gravestone, your enshrined name woven in with the characters that signify that you were once a soldier, and even in death, you remain as such. Your family will only be able to talk about you in the darkest hours of the night when no one is around, whispering the smallest details that make you a singular human being, a unique member of a family, amongst themselves like a secret. What a small price to pay for the honor.

May 1939: Shirō Standing Guard, Walking around Jining

Guards, always facing the the vast landscape, scanning the terrain for any suspicious movement. Guards, guarding the fortress town at every hour. From what? The Nationalists? The Communists? The wind, someone says, is carrying the desert sand. The wind, someone says, comes from afar, its path unblocked. All around the fortress town, wheat fields that stretch from one end of the horizon to another. Not good for farming, Shirō thinks absentmindedly as he stands guard atop the fortress. At home, they must have opened the irrigation to let the water into the rice paddy, he thinks, and he can see the land around his house turning dark brown from the water, then submerging for the season under the mud. He thinks of what to write home, because they expect him to write back of all the things he sees; he knows that he owes them that much: that in this fortress town, there is a curfew. No. That there is a market, but it is a sad affair, too small to be called a market. No. That food is cheap. Yes. That there is a comfort station where his brother-soldiers go at night for quick and cheap sex. No. That the fortress town is riddled with the memory of battles fought, walls still stained from blood, bullet holes at chest level on the wall such that whoever was between the wall and the rifles must surely have died. No. That the Chinese are inscrutable. Yes. That he is fighting well. Yes. That he is doing well. Yes. That he thinks of home, and he thinks of his wife, Masa. Yes. Always.

May 25, 1939: A Letter from Shirō to Masa

My dearest Wife, Masako,

We didn't have chance to talk alone when I was on the leave but I'm sure you understand what I am thinking. It was such a wonderful time. I'm so happy to see Seijurō growing up to be a wonderful boy. With the boat engine, every sound was erased, but I can still vividly see your gentle smile as you saw me off. Nothing special happened after we left Shibaura Harbor and for a week, and we landed in picturesque —— Harbor, spent the night there, and we arrived to where we are on a train. China is a continent, but we called it an Old Big Country, and I see that as I am here.

The first thing I noticed when I landed was many coolies — day laborers who make their living sweating and from day to day. For the first time, I was so happy that I was born in Japan. We spent a night at —— and got on the train for two days and two nights. They said that the train track was dangerous, so I slept with a loaded gun under my head. There were many Japanese living in towns along the railroad track and they came out holding Japanese flags in their hands; women were dressed up and welcomed us. I was very impressed. As I wrote before, there are never-ending wheat fields as far as eyes can see. The current town ——'s population is about 130,000, so I would say that it's about the size of Sawara. All around us there are two, three walls made out of stones and no one can enter from outside; on all corners of the fortress, Japanese soldiers stand guard.

When I think about this fortressed town and a company guarding and suppressing the rebellious Nationalist soldiers, I think of the seriousness of our duty. Sometimes, we guard the envoys going to a town sixty kilometers away, and have to break through the enemy line. Like I wrote last time, the weather is like the middle of the summer in our homeland. There's no rain and like I wrote, I'm doing very well. I am doing well. I haven't gotten sick at all. We finally found water for the supplies and I'm also used to it.

In China, they work at their own pace and they also have physical endurance and they know how to be two-faced. Japanese had to find this out the hard way. In a strange way, they can be cute like children, when they come over, calling us "Sir, Sir." When Chinese children see us, they come pestering, saying, "Give me cigarettes, give me cigarettes" — some are cute — they remind me of Seijurō. Selling and buying is the Chinese way, with the prices are doubled knowing that it's going to be bargained. One beer for 90 *sen;* cider for 30 *sen.* A bit too much, if you ask me. In general, sundry goods are expensive, but it doesn't bother me. We have alcohol coupons, so I can buy it cheaper than back home. Chinese people have a four-thousand-year history of kowtowing to the ones who they think are stronger and looking down at people who they think are weak, so I cannot relax even for a minute. Right now, they kowtow whenever they see Japanese and pretend to be friendly to us, but I've seen what they are capable of doing. There's something very eerie about them.

Also amongst the population of 130,000, I have yet to see Chinese girls of age; they have a strong sense of purity and virginity. They all hide in their houses. We are currently staying at a school and a factory that makes Chinese woks, so we live amongst the Chinese. Everything they do is according to the Japanese military style. We work at bridging the relationship between Chinese and Japanese. Of course, the school is taught by a Japanese.

From tomorrow, I will be guarding as well as beginning a new duty. I thought I'd go see my cousin, ———. I've gone to the town of ———, but since I didn't remember the name of his regiment, I couldn't find him. Also, I couldn't remember the name of his wife, so I will not be sending a letter to them. Can you do this for me?

The golden wheat fields as far as my eyes can see is now about to be harvested. I imagine that back home, you are all busy with the field, but since you have a weak constitution, do look after your health. Please honor and serve Grandmother and my parents for me, because I cannot. This is what I wish the most.

And since Seijurō is of the age when he is getting smarter and smarter, please educate him with your intelligence. I am doing so well physically. I grew a beard and it was getting to be better looking than my captain's, but regretfully, before I had a chance to get a photo taken, it was shaved off.

We have electric light in the room and there's nothing we are lacking, but since this is the frontline, it's not the same as being at home. There are some funny men in the company and they entertain us. This is where Confucius was born. I will write more later, about interesting things and details. Send my regards to my sisters and your home.

Goodbye
Shirō
China
Yamaguchi Company, Takenoshita Battalion, Watanabe Squad

June 10–26, 1939: The First Battle

Cars stretch forth on yellow earth, trucks carrying men after men dressed in khaki, yet the colors do not hide them in the landscape. They each carry the flag of the rising sun; the truck and car each carry the children of the Emperor to eradicate the guerrilla Chinese, who may be hiding anywhere, everywhere, who may pretend themselves to be *good citizens,* even carrying the pass to prove themselves innocent, but you can always tell who is a Communist by their hands: soft hands of an intellectual just learning to carry a rifle. By their forehead: the tan line on the forehead that demarks where the face ended and the hat started, another recent exposure to the sun. Shirō is in the cabin of the truck, because he is a noncommissioned officer, and he looks behind him to see that his men are still there, all huddling on the truck like scared hens. Some smoking to hide their fear. Then suddenly, a shrill shot particular to the Chinese rifle, *Guerrillas,* someone yells from ahead, and they all scramble from trucks and cars, they run as quickly as they can, stumbling from all the gear on their backs, tripping over each other, trying to take cover. They all huddle behind the truck, its engine still running, pushing each other for a deeper cover, the side is riddled with bullets, and the rifles start shooting from this side, from that side, the metallic shrill deafening, he hears his officer yelling, *Go, go, advance,* and Shirō yells to his men, *Advance, advance,* but no one moves, and he can hardly move himself, but different officers yell, *Advance, take cover, retreat, advance,* and his own body moves forward, takes cover, retreats, forward, just like the commands, just like he was trained in the basic training, the body moving automatically to the order. His officer yells, *there, there,* and he sees it: a handful of enemies behind the shack, shooting, and he pulls one of his men by the lapel, he moves away from the cover, he slaps obedience into his men and yells at them to advance, advance, as he rushes forward, leading his men, half crouching, half scared, but rushing with battle fever, toward the enemy. This is only the start of the eradication of the guerrillas. They march. They keep marching. They take one fortress after another, killing everything that threatens them, raising the flag of the Rising Sun over every ruin.

June 1939: What a Soldier Takes with Him into the Battle in the Private Item Pouch, Which Hangs Round His Neck, Right against His Heart

— Photos of his family
— Talismans of various kinds to keep themselves alive: a little handmade doll, Sennin Bari, or the Thousand Stitches, a talisman pouch from the local shrine
— Military notebook with their blood type, height, home address, military status
— Military-issued journal book to write their thoughts down
— Pens or pencils
— Postal saving book and the family *hanko*
— A folded-up Japanese flag with people's well-wishing messages

July 3, 1939: The Fall of Heze

For nearly a month they have been marching, after they lost their trucks. After the rain when the trucks were rendered useless, and they had to rely on their own legs to carry themselves forward, with all their gear on their backs and no one to carry it for them. They have slept on the road, in the rain for a month; they carry themselves forward. Flies cover their backs, attracted to the unwashed bodies. Water trucks follow the march like a herd of elephants, stopping every hour, hour and a half, to moisten the perched lips, cracking and bloodying from too much licking, biting of dry skin, but the water is not sanitized enough, and men bend over in pain after a few minutes, running toward the edge of the road to pull down their pants. Men do not have enough time to write letters home, they don't have time to shit properly either, but it doesn't matter. What they took as a leisurely activity back home is no longer something to savor, but something to do quickly; as they march, one veers left, pulls down his pants, and shoots out liquid, then even without wiping, he pulls up his pants and keeps marching. Most limp, because the leather boots do not fit their feet, and no matter how many layers of socks they wear, their feet blister, burst, making each and every step they take an agony, and when they do take off their boots, a sheet of the skin of their feet comes off with the rotten socks. But here they are, finally: Heze. The last battle. The red zone. To conquer the fortress of Heze is to have won the battle. The last fortress. The battle is fierce: men fall like slapped mosquitoes, men falling, yelling for help, bullets flying, and they keep shooting, shooting, mortars exploding this way, that way, men are shitting in their pants, screaming, battle fever overwhelming them, firing erratically without aiming, the barrels of rifles too hot to aim straight but they pull the trigger, then reload, pull the trigger, then reload, using up all their bullets, and when the bullets are gone, they fire their semi-automatic handguns. All the hand-to-hand combat training: useless. The enemy finally surrenders, walks out with a white flag. They stop firing, finally, after yells of *cease fire, cease fire,* and capture these weak-willed soldiers, tying the hands of the prisoners-of-wars behind their backs and stringing them on one rope. The flag of the Rising Sun goes up. They cheer, *Long Live the Emperor! Long Live the Emperor! Long Live Imperial Japan!,* and the cheer echoes in the clear blue sky, but the dry earth sucks it up, just like it has sucked up the blood of the fallen, both the friend and the foe.

How the Bodies of Our Brother-Soldiers Are Collected

Because we are human, we do not disappear like cats into death, alone to where only they know where to go. We want our last moments to be noted, to be remembered by others. We cannot care for our own bodies once we are dead, so we honor the bodies of our brother-soldiers in the way we want to be treated. When someone dies in the hospital for whatever illness, whatever injury, this is easy. There are people whose job it is to note, to document, to cremate the body, pour the ashes in an urn and ship it home in a box. But the fallen brother-soldiers in battles, we will do our best to gather them. If we are at the very edge of the frontline, if there is no time to gather the dead, to send them back whole, we crawl to where they are and cut off body parts: arms if they are officers, fingers if they are infantrymen. Amidst mud, bullets, and our own fear of dying alone, we carry fingers in empty cans of toothpaste as we push the line forward. See, there's a company monk with a donkey trailing behind with five arms dangling, names of the dead beautifully written as not to mistake one officer from another. We fear dying alone in a strange land, so far away from home. We fear being left alone, not quite dying, in a strange land where no one speaks our language, where no one knows who we are. We can always tell how war is faring by how we treat the dead. When we must leave the dead behind, that is when war has turned uglier than we ever imagined possible, when war is no longer about winning but about retreating, about surviving, about leaving behind the dying and the wounded to fend for themselves, giving them a hand grenade, just in case. That is when we know we will lose the war, but we must keep going because there is no way back, at the same time as there is no way forward.

Soldiers

These are men with family, with pasts, their dreams they have not yet dreamed and hopes that have not yet actualized. Who have loved and who have grieved. Men who have lived in cities; men who bought whores; men who are virgins; men who beat up their wives and children; men who show tenderness. Now, at the command of the State, in the name of the Emperor-god, they must stop their daily lives, their lives interrupted in mid-stride. They are suddenly victims because their fates are suddenly intertwined, entangled with the fate of the nation, and they must leave behind families without *daikoku-bashira,* the breadwinner. Sons must obey their fathers, and the Emperor–god commands them to fight. They go through basic training, where each and every day their personal lives are rendered meaningless, their lives become public, collective, modeled after the imaginary *soldier,* the ideal created out of an improbable nostalgia. The Father commands, and it did not matter that these men refused to *kill* at first. They all do. But as they live through battles, watching their *brother-soldiers'* fragmented body parts flying through the air, as they listen to their *brother-soldiers* crying for help, for their mothers, wailing beastly cries, as they see horses they so loved lame and they are the ones who have to pull the trigger — and they might have to do that later for their brother-soldiers, as they are commanded to *disable the enemy* but they are killing women and children, their hearts harden. They do kill, and after the first one, it becomes easy. It is easy. They have turned into soldiers. A collective force, acting as one, thinking as one, killing as one, anonymous in their uniform. They will try to keep a part of themselves private, they will try to keep their souls intact through letters and journals and photos they carry close to their hearts. They are soldiers because they want to go home alive. As in life, these men are not one-dimensional, but must negotiate between contradictions.

The Infantry Is About Walking, Marching, Walking

They carry everything they need in their backpacks, clothing, personal items, bullets, tents, everything they need to make their lives bearable, a semblance of normality, even though they are in a foreign land, in hostile territory, fighting a war that seems to have no end, marching on land that has no end, surrounded by people who can be their enemy or not, speaking in a language they don't understand. When a common language is reduced to a mere five words, *thank you, give, whatever will be will be, girl, goodbye,* there is no knowing the other. Twenty kilograms of sentimentality and necessity, almost half their own body weight. And flies that cover all exposed flesh. They walk. They have been walking for such a long time that they can sleep while they walk. *Pissing takes 109 m, shitting takes 872 m, and raping 2180 m.* Don't fall behind, even while you're pissing. And they cannot falter, for if they fall behind, who is there to pick them up? Who is there to protect them?

August 2, 1939: Avenging the Death of the Company Commander

Trucks rev up the engine. Cars behind them rev up as if in reply. The mission: to avenge the death of the Company Commander, who fell under an enemy bullet. No mercy must be shown, it is understood. No compassion to the enemy. The soldiers are ready; Shirō yells at them to get the battle fever going. The first truck rolls forward, filled in the back with soldiers armed with bayonets attached to their rifles, hand grenades hanging from their belts for easy access. The mortars are ready too, gleaming black. They move as one away from the Heze gate. Nothing will remain standing in their path. Everything will be burned down to the ground.

How the War Dead Come Home

A soldier of the same rank as the dead carries a white cloth box hanging from his neck. There may be twenty. There can be hundreds. It all depends on how many battles were lost in the past three months. These boys left six months ago, amidst fanfare, with every woman and child waiving Japanese flags, these soldiers marching in step with music. Proud sons. Proud and chosen boys. Look how beautiful they were — they were the best of Japanese boys. So young. So faithful. The Emperor's children, off to the sacred war. The holy war. And now, they have come back fallen. But no, they have not really died, but are on their way to the Yasukuni Shrine. Now, they are going to be gods — they don't have to wait the usual forty-nine years. After they have been praised and grieved by the regimental headquarters, they are carried home to their respective villages, and each village carries out a funeral fit for a god. After all, these boys left the village, chosen out of many, to represent the village, to bring honor to the village that raised them well. The entire village grieves for their prodigal sons. See how all have shown up in their very best. The entire village has come to the town hall to honor the passing of a new god. Now that the state has honored these boys, ash in the box is returned to the family, and these families are now called *homare no ie — the Revered Home.* There is no greater honor for a family to have than to have a god amongst them, though he is not theirs to claim.

August 23–September 15, 1939: An Easy War

Battles one after another, as if it is one long battle that has been going for nearly half a month, with small pockets of time between fighting. Yet, Shirō and his men have been moving, marching forward, and every exchange of fire a victory. Every village eradicated, leaving behind wailing women and children, and, sometimes, dead women and children, though most of the time, they try, they try so hard so try not to shoot them. But what happens when a woman takes a semi-automatic machine gun and starts shooting? What happens when a child totters over with a hand grenade in his hand? No one is to be trusted. Not Chinese men. Not women and children. No one except for their own brother-soldiers who would do anything to save them, anything, even throwing themselves in front of others so that others can live. But now, they are tired. They have not drunk decent water for a long time, and the only thing that they can safely drink is alcohol, and there is enough of that to numb the exhaustion, to numb what they are doing. Marching. Sleeping on the hard cold ground. Eating preserved food, nothing fresh except for what they take from the peasants, who plead with them not to take their harvest away, who plead, getting on their knees, begging, begging, but they slam the butts of rifles against their bodies and turn away with cabbages, green onions under their arms. Of course they feel guilty, especially the ones who left behind a farm back home. But they tell themselves, *there's nothing I can do.* And this is what they must do for an easy war where they leave behind houses smoldering with smoke, farmlands stripped of the season's labor, and the wailing of men and women.

September 1, 1939: The Day of Service

Today is the day to celebrate and give thanks to our soldiers fighting the Holy War in China. China, the child-like country that spits back at its big brother, Japan. We are here to lead the whole of Asia to fight against the Western imperialists, those who want, want, want but are not willing to share. See how many nations have fallen under the greedy white men; Japan is the only country that has resisted the Europeans and Americans. We dream and work for Great Asian Co-Prosperity Sphere, and today is the day to show our gratitude to our sacred soldiers fighting the Sacred War. Today, every meal will consist of one side dish and one bowl of soup. There will be no smoking or drinking, and all restaurants and amusement and recreation establishments will close. We must feel what our great soldiers are feeling. We must not dwell in luxury, when our soldiers are fighting gallantly on the harsh terrain of China against the unyielding Chiang Kai-shek. And from now on, every first day of the month will be Service Day.

September 1939: Understanding China

Shirō stands atop the fortress wall, scanning the land all around, the solid gold of wheat fields with small shack-like houses dotting the terrain. So unlike his home, where the earth is dark brown and water plentiful, the river behind the house clear and sweet to the tongue, and where now the golden landscape must already be waving, bowing with the wind, the shape of the wind made visible. In his memory, home is beautified, molded into mythical proportions, and the present land, incomprehensible. He does not understand the language of the land. He does not understand the inscrutable faces of the Chinese. He does not understand the rules of this nation, both understood and unspoken. All that he knows of this land is through the filter of fear and battles, through the rhetoric of his own nation, spoken by his commanding officers and hearsays: *the Chinks are a bunch of peasants, they don't know what they are doing, they breed like rabbits, they don't mind living in dirt and shit and with pigs.* That is all he knows and understands.

IMAGINARY DEATH

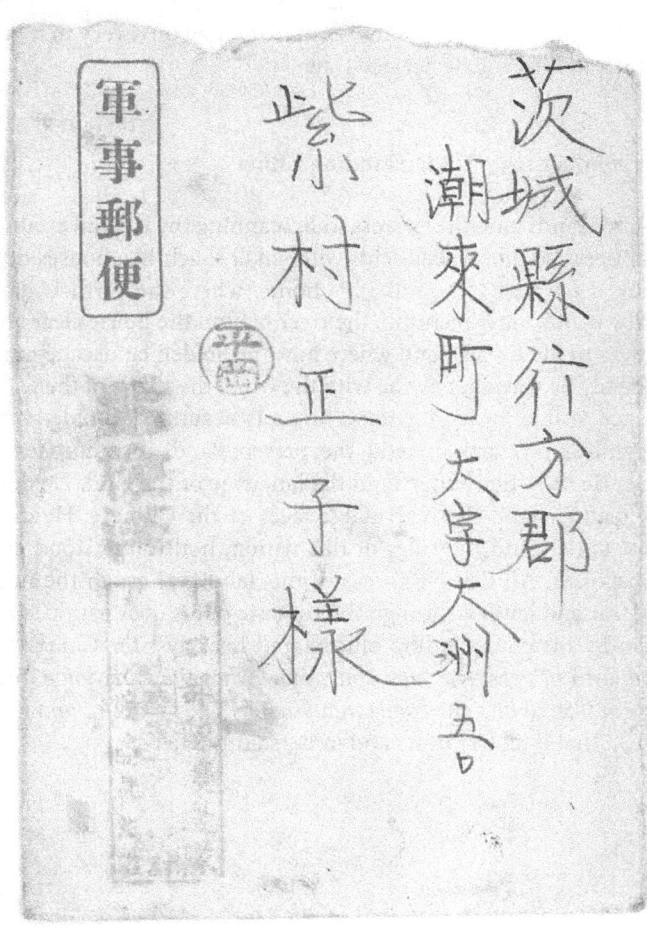

Fig. 3.2. Letter from China, envelope (front).

THE SACRED WAR

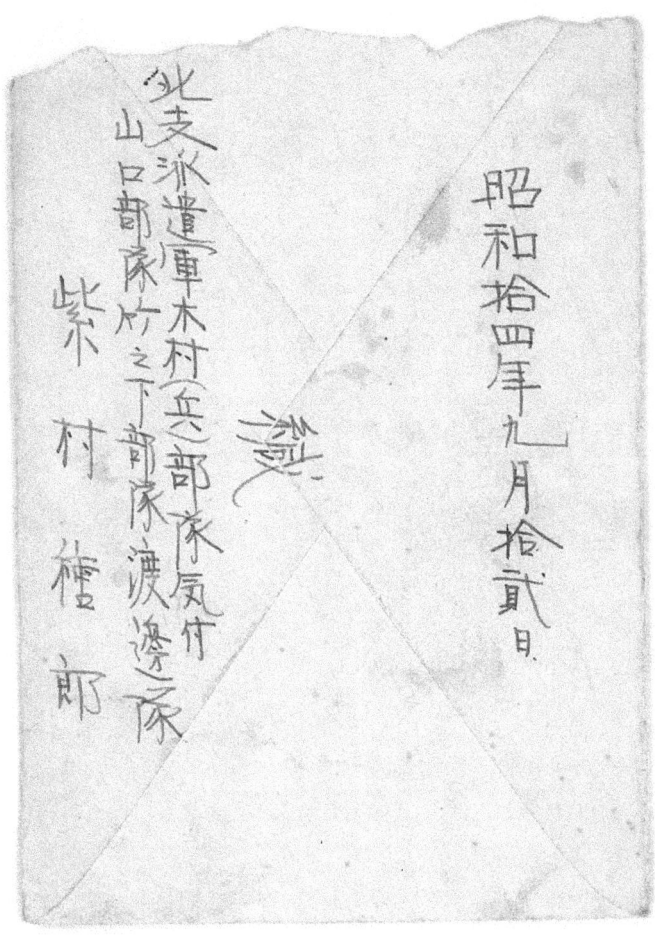

Fig. 3.3. Letter from China, envelope (back).

IMAGINARY DEATH

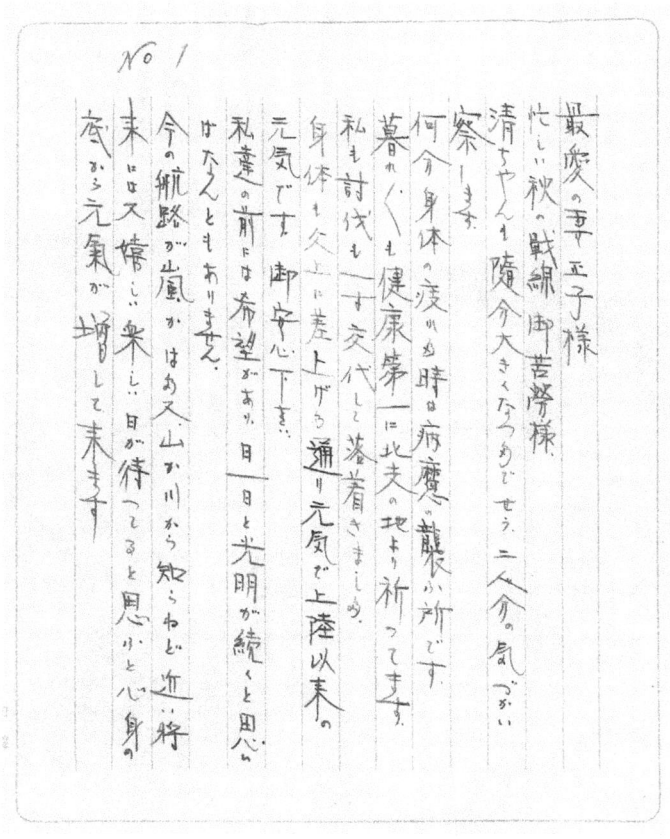

Fig. 3.4. Letter from China, page 1.

THE SACRED WAR

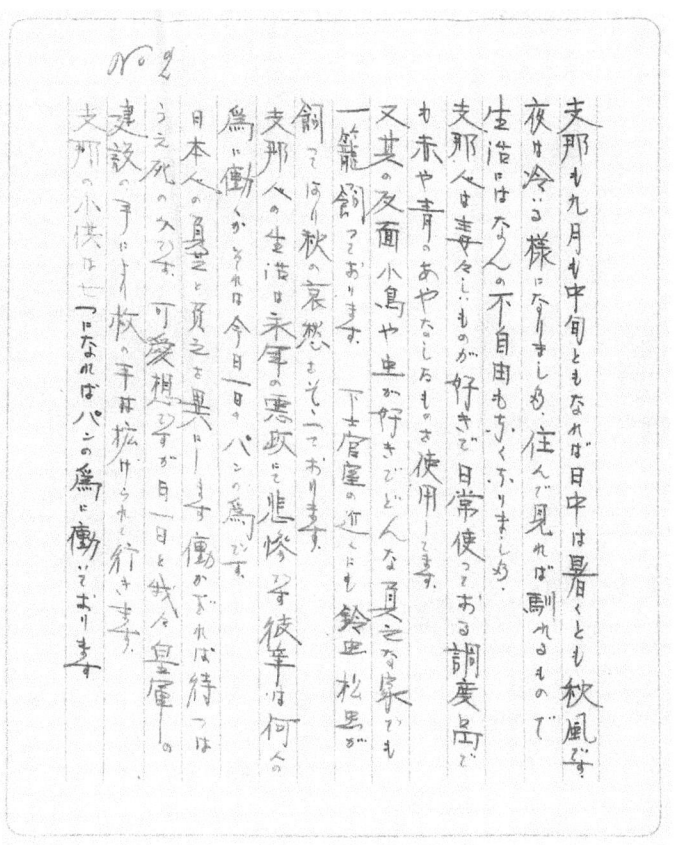

Fig. 3.5. Letter from China, page 2.

IMAGINARY DEATH

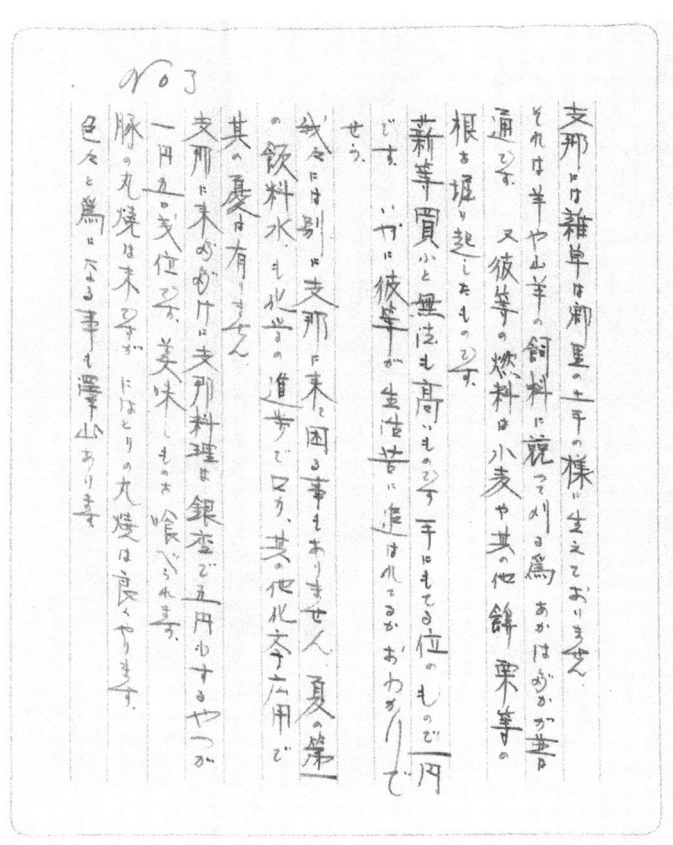

Fig. 3.6. Letter from China, page 3.

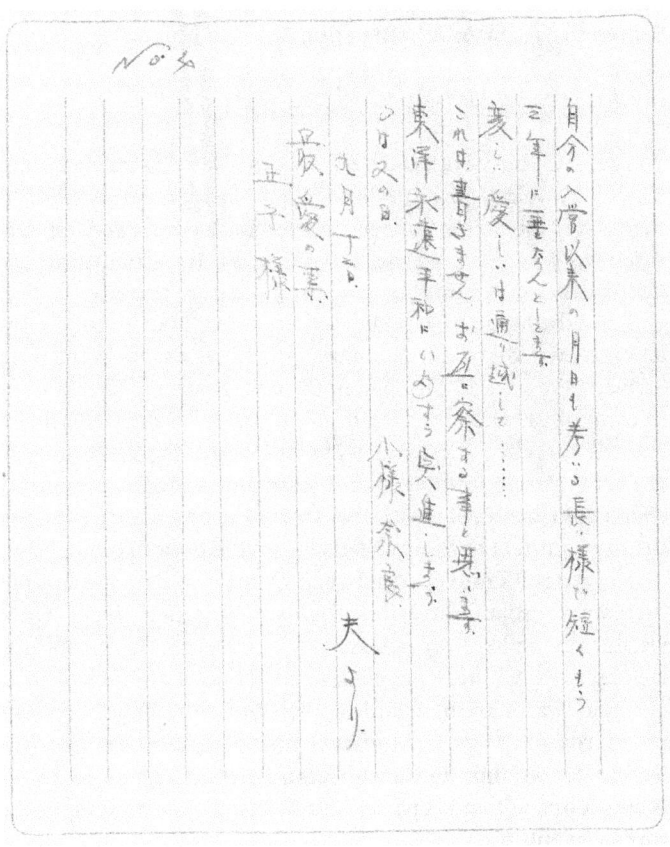

Fig. 3.7. Letter from China, page 4.

September 13, 1939: A Letter from Shirō to Masa

My dearest wife, Masako,

You must be busy with the harvest — the war that farmers must face every year. I imagine little Seijurō has grown so much. I thank you for taking care of him for both of us. You get sick whenever you push yourself too much. I pray from northern China that you are putting your health as your priority.

I and the battle have finally settled down. Like I wrote in the letter to my father, I am doing well, exactly the same as the day I arrived here back in May. Do not worry about me. There's hope in front of us; when we keep in mind that there's day and night, and the sun always rises after the long night, there's nothing to it. In my current journey, there's storm and mountain and river ahead but when I tell myself that there's happy and joyful days in the near future, I become cheerful from the bottom of my heart.

Now that it's mid-September in China, the days are hot but there's also an autumn wind. Nights have gotten colder. Once we start living here, we get used to it, and now, there's nothing lacking here. Chinese people like gaudy colors and even their everyday wares are blue and red. On the other hand, they like little birds and bugs, so no matter how poor they are, there's at least one in a family. In the noncommissioned officers' barracks, there's a house that keeps crickets; it brings on the song of the sorrow of autumn.

The lives of Chinese people are heartbreaking due to a long reign of cruel governing. What and who do they work for? For a day's bread only. Their poverty is different from Japanese poverty. If they do not work, the only thing waiting for them is death through starvation. They are pitiful, but their lives are getting better and better due to the Imperial Army's help.

Once Chinese children are seven years old, they begin to work for bread. In China, there are no reeds like the ones on the river banks back home. It's because they chop them off to feed sheep and goats. And their fuels are made out of dug-out roots of wheat and rice and millet. It's very expensive to buy wood for fuel. An armful would cost one *yen*. You can see how much they suffer just to stay alive. There's nothing inconvenient here in China. Thanks to the advancement of technology, we don't have to worry about water. There's nothing we need to worry about. You can get, for one *yen* and fifty *sen* here in China, the kind of Chinese food you'd pay five *yen* for in Ginza. There are so many delicious dishes. I haven't had the chance to eat a whole roasted pork, but I've eaten roasted chicken quite often. There's much to learn also.

It's been nearly three years since I entered the service; it feels like such a long time, but also such a short time. I've gone beyond missing you or yearning for you.... I'll stop here. I'm not going to write this. We both know what I am feeling. I will keep going forward for the eternal peace in Asia. I will write more later.

Goodbye
Your husband

What Makes an Imperial Soldier in the Sacred War

He — after all this is the time when men are men and women must stay home — would be carrying a rifle known as a 99-Model which has to be cocked every time he fires, five rounds in total. Remember, it carries the holy insignia of chrysanthemum. The Emperor has lent it to him while he is a soldier. To lose it, to break it, not to clean it, is to spit back at the Emperor–god. Do not disgrace the rifle. Do not disgrace the name of the Emperor by breaking it. If his squad leader is the by-the-book kind of man — and they are always that kind of man — he will make sure each soldier has 120 rounds of bullets at all times. Not 121 rounds, not 119 rounds. If there's one missing, the soldier will do anything — even steal from another squad — to make sure he has exactly 120 rounds. The bayonet is the most important thing, after the rifle. Do not forget to clean it. That is the extension of the rifle; that is what makes the man a real man: the rifle is the body of the emperor, and the bayonet, the arm of the soldier. Close combat is what makes a real soldier, after all, not the mortars. Not the hiding. The first and final test after the basic training will be how to thrust a bayonet, at the right angle, with the blade flattened enough to slip between the ribcage, and the twist. That's what he will have to do at the very end of the basic training, if you're in China and there's enough Chinese POW's around. Never thrust a man in the stomach. It is messy, that's what he'll do if he's scared shitless, and he will be. Remember, it takes seven seconds for a hand grenade to blow up. Once he pulls the pin, count to three, and then throw it at the target. If he throws any earlier, the enemy will throw it back at him. Any later, it will blow up in midair. But do not use the entire supply of grenades; each soldier must save one, just in case. The Emperor knows that soldiers are fighting a holy war and in a holy war no one dies. Soldiers become gods in death. But in case, just in case, there might be a circumstance when a soldier might have to sacrifice himself so that the others can keep fighting. Keep a grenade. Do not forget to recite the *Rescript for Soldiers and Sailors;* these are the divine words given to each soldier by the Emperor. His words must be followed; besides, an Imperial soldier does not fight based on material resources. He fights with his spirit. Like all the great warriors of the past who fought under the most disadvantageous circumstances, all you need is the fighting spirit.

September 1939: The Reason for Fighting

Shirō understands the reason in a way he understands something far away, something that he has read in history books about the ancient past or in books about dinosaurs. He understands this fight, this battle. The reason that he is here is for democracy, for freedom, to protect these Chinese people from the tyranny of the State, for Japan, the leader of Asia; he must teach these Chinese peasants what it means to live in peace. He does not question. He fights for a greater cause: *It's not the fault of the people, it's the government, it's Chiang Kai-shek who must be punished.* But why is it that the faces he sees as he walks in town are sullen and angry; why is it that young girls run away as quickly as they see him or his men; why is it that vendors, in their private moments, when they think no one is watching, spit in disgust on the dry ground, raising dusty clouds? He knows he must fight, that he is serving his country, and to him, questioning is not a part of his training. It cannot exist, because once he does, all will shift and redefine itself to create a new world in his mind. And he cannot allow that.

September 30, 1939: *Give Birth, Produce, It's All for the Country*

Get married as soon as possible; give back to the country by offering children to the Emperor–god. Who should bear children, who cannot, is decided by the State, just as the Yasukuni Shrine decides who are the war dead and who aren't. And for bearing children, the State awards those generously, for the country wants the future to go, for the Emperor–god to have more children. And the birthrate goes up. Five hundred forty-four thousand men are drafted this year. The State is pleased. The Emperor–god is pleased.

October 1939: The Soldier's Heart

It is always the left side, the heart that palpitates at unexpected moments, the sudden exhaustion that feels like Shirō's body is suddenly weighed down by gravity, or the brevity of it all that he is not allowed to talk about. Shooting. Killing. Fighting. So instead, his body betrays him, or, his body tells the truth because he cannot. The sudden stabbing pain in his heart. The shortness of breath. The dark moments between gun shots where he feels, he knows, that he will not come out of this alive. He smokes one cigarette after another trying to ease the pain, or for company,

IMAGINARY DEATH

Fig. 3.8. Shirō in China.

or for ... he does not know why he smokes so much, even after he had promised Masa he would cut down on how much he smokes. The only thing that keeps him going is the batch of old letters he carries next to his heart, where it aches, and photographs of his family. He knows that these are the only things that keep him going.

December 10–20, 1939: Shirō on His First Return to Japan

The distance between the battle and home, only a mere week, yet how different it is, another country of his imagination. As the ship moves away from Qingdao harbor, he looks at the red roofs, the European streets, the unyielding country so stubborn, the harbor so different from the day he arrived seven months ago. The country he saw seven months ago. The country he sees, leaving it behind for ten days for an assignment in the motherland: to pick up fifty newly trained men and bring them back to Heze. Yet, as he looks out of the window on the train from Shibaura to Sakura, where the 212th Regimental Unit Headquarters are, it is as if the war has not touched Japan, as if there is no war anywhere in the world. The food is bountiful, with the lunch boxes still filled to the rim with white rice. Men and women laughing on the train as if there is no care in the world. Only once in a while, there are soldiers on leave, or perhaps veterans in their khaki uniforms, who salute when they see Shirō with his sergeant's insignia. With two days left before he leaves, he makes his way toward home, only an hour away from Sakura, but the leave is too short. His son does not remember who he is, no matter how many times Masa pushes him forward toward Shirō; he so wants that minute, that hour, just to be alone with Masa, but people keep interrupting them, wanting to congratulate him, as if by associating with him, they, too, are basking in the heroism of war. Masa, Masa, he thinks, and they keep making eye contact amidst the bustle of people celebrating his short homecoming. And two days later, he is preparing to lead his new men down the boulevard of Sakura, to take these men back to China. It only takes seven days to leave the front. It only takes seven days to go back to China, to forever change these young boys who are excited, who are going to be the narrative of the nation, some who may not come back alive, though they do not know this yet.

1940: Year Two Thousand Six Hundred

Years are measured by politics, by emperors' reigns. Each year signifies the year of the Emperor's reign. This year is Showa 15. It is 2600th year after Emperor Jinmu, the direct descendent of the Sun Goddess, Amaterasu Ōmikami, founded Japan. Amaterasu, the divine daughter of Izanami and Izanagi, the brother and sister couple who experienced the first celestial marriage and the first celestial death. The goddess came down to earth and lived amongst men, procreating on her own and out of it came the first Emperor. No one says that this is a myth. Here is another reason to drown in the year-long carnival amidst drums and dancing.

Yasukuni Shrine, the Beatification of the Dead, the Abolishment of the Individual

The war dead are no longer individuals with singular histories, with details and deeds they regret, loves that broke their hearts and dreams they dared to make real. They are gods, sacred, devoid of humanness, only receiving the praise befitted for a god. Now that they are gods, they take on the same story: they were good sons, they were earnest and diligent students, and they supported their family by waking up at 4 a.m. to do the farm work before walking five kilometers to school. They so wanted to go to high school but they gave it up for their younger siblings. And they are good sons now that they are dead, having sacrificed their lives for the nation. They leave behind mothers who are glad their sons died honorably; they leave behind wives who teach their children to avenge their fathers' deaths, who will offer sons back to the Emperor–god. But before they can ascend to godhood, the most private act of living must be taken away. The living cannot grieve; the war dead are carried through the public ritual of the village funeral, the public mourning, medals awarded and eulogies made. They are given a new name — *eirei* — the heroic dead, a bind that instructs the bereaved how to conduct themselves; it also binds the dead in how they must act, even in the otherworld.

February 9, 1940: Shirō, Writing a Letter to the Hara Family

He now knows that no one dies yelling out, *Long Live the Emperor!* No one dies heroically the way they write in newspapers and magazines; no one dies quickly in battle or on the battlefield. Most die calling out the names of their mothers or their wives. Some that died a lingering death from a wound or sickness left incomprehensible words, words he could have deciphered if he were a better man. Yes, he tells himself, if he were a better man, a better commander, he would have understood what regrets these men left behind; now, they would remain unfathomable, like the will of the gods. On February 9, he lost Private Hara to a lingering sickness. They cremated him at the base; the ashes were placed in the squad hall, and the soldiers all grieved for the loss of their brother-soldier. *He has become one of many, many war dead. Go to the Yasukuni, be a god there, show us — the squad brothers you left behind — how to be real soldiers.* After his death, he has been raised to the rank of Private First Class. The family he left behind will be taken care of, with the pension befitting his rank, though it will not be much, not enough to keep the family alive. Now, Shirō will have to write a letter to his family, ignoring how Hara had suffered, how he just wanted to go home, how he was sick for a long time. Hara was a young boy; he died before his first battle. It was not an honorable death, and Hara had been ashamed. The last time Shirō went to see Hara was two days ago, when Hara was in critical condition. Writing this letter is the least he can do for the memory of Hara; it is the least he can do for the family. It is better they do not know the truth; it is better if they go to their own graves believing that their son died heroically. Shirō knows this much: Hara will become a god at the Yasukuni; he will look after this company. Hara did die honorably. Shirō takes up his fountain pen and starts: *I am so sorry to write this letter....*

Those Who Die Before Their Parents Must Languish in Hell

It is said that children who die before their parents can never go to heaven, but instead are punished for causing profound sorrow to their parents. Punishment: to wake up every morning in hell, and they must go to the riverbank and stack up rocks high enough, solid enough so they can climb up to heaven. Of course, there's one thing: a herd of devil creatures would come out at sunset and topple over their labor. The children have to start all over again in the morning, all through eternity and more. During the war, this idea of hell is suspended. Buddhism — the godless religion — has been pushed aside by a living god. And this living god — and all the gods before him — has decreed that a child is born of parents, but the ultimate Father is the Emperor, and for that, there is no hell if a man dies in war; a man becomes a god.

March 23–May 12, 1940: Taking the New Recruits to Battle

They are not ready, Shirō knows that, but as an officer who has experienced battles, no one is ready until you experience your first battle. And the second. And the third. And before you know it, you have learned to cope, to distinguish which shrill belongs to the friendly fire and which to the enemy's. And even the mortar shells that may fall, you learn to distinguish the distance by the sound, just as you would on a field with thunder. But this is how a recruit becomes a real soldier, when all that they've learned can be put into practice, and all the target shooting, all the bayonet training, can be put into practice. The enemy is not like a target practice: the enemy will keep moving, and these new recruits must be able to foresee where the enemy is going, to learn to pull the trigger before the next move of the enemy. All this. They keep marching, marching, and no one has time to write letters, and Shirō does not keep a journal, not at this time. And perhaps the rest of the time during his stay in this heathen land. He does not record how he feels about receiving the Order of the Rising Sun White Paulownia Medal. Just because it is the 2600th Year, just because he has been serving for long, just because he has survived this long. Maybe he is proud. Maybe he feels ashamed that he is still alive while men he trained have fallen, that maybe he did not instill enough fear in these men to obey his orders. Maybe he does not know why he received it, but thinks that the Emperor has chosen him. We will never know.

1940: A Remembered Photo

Look: there is Shirō, somewhere in northern China, standing, staring at the camera, with a full beard. The recollection of the photo occurs sometimes after the war: Yoshiko remembers her mother opening the drawer, a private moment between a mother and a daughter, two conspirators sharing a secret past. Yoshiko remembers her mother, so many years after his death, whispering to her, *Your real father told me that in China, no one took you seriously if you were a man without a beard, see your father with a beard here? He sported a bearded when he was in China.* Beard, in the land of the wise old men, symbolizes power and wisdom. This is why he grew a beard, Yoshiko told herself, and she holds that knowledge of China even now: in China, you must grow a beard. Shirō stands in front of the camera, with his back straight and his sword placed in front of him. He is an Imperial soldier, a child of the god, and he is there to liberate the Chinese from the tyranny of the state. He is in a holy war. But this is a remembered photo. A photo lost somewhere in the past, lost when his wife died half a century later.

Fathers

Authority lies with the father. What the father commands, the rest obeys without question. His words are the words of god, and his family, his faithful believers. We do not seek love from the father, we seek his protection, his approval, and he will look after us. Before the Emperor became a god, the authority lay with the village, and once He was brought out from obscurity and rose to the height of godhood, *living god,* He became our Father. And a good subject is the one who serves Him obediently. We are all His children. Father commands us to fight and we now fight for the god, and because we fight for and under the god, we fight a holy war.

April 19, 1940: A Letter from Shirō to Masa

Dear Masako,

I've been meaning to take up my pen but it's very hard to do on the battlefront. One battle after another, another battle then another one; it doesn't give us a luxury to pick up a pen. And days pass.

I received a letter from Father saying that everyone is doing well. I am relieved. Though I am busy, I am doing well. Don't worry — I'm keeping up with everyone else. Here in northern China, summer-like weather — so much like summer back home — has already arrived. The wide, big wheat field is full of ears. It seems like there's so much news about the northern China in the Chiba version of the newspaper, but I bet there's nothing in the Ibaraki version. In the last series of battles, we've lost Captain — as well as — men. Luckily, I am doing well and am fighting hard to avenge these men, so do not worry about me. I've been meaning to send you photos but I haven't been able to. I just got two photos, so I am sending them to you. I will write more details later.

From Shirō

I wrote a letter to Father, so have him show it to you.

How the Fallen Bodies are Collected

Shells shrill through the air. Bullets *pop, pop, pop* left and right. A beastly scream. Another brother-soldier stops in mid-air, and without a sound, crumbles, as if the strings that held him up were cut abruptly. The enemy we cannot see. Bullets that appear from nowhere. Shells exploding suddenly next to us, left. Right. The battle lasts a lifetime. And no time. Time is as elastic as the body in midst of the battle. And after the bloodshed, there is much we need to do: we must go from one fallen body to another, and if we find the dead, we must carry the bodies back to the rear if we can, but if we are at the front, we cannot leave the bodies the way they are for the others to retrieve them. Chinese are ruthless; we have seen what they have done to our brother-soldiers. Instead of dishonor that the dead may face, we cut off the body parts to carry with us: arms if they are our officers; small fingers if we are at the foremost in the battle line. We will carry the parts of the dead with us. We cannot take care of our own bodies once we die, so we honor the bodies of the brother-soldiers in a way we want care to be taken of us.

Gods Going to War

The gods all over the land ready for the war, and they go where the faithful go. And it is easy to keep track of where the faithful are, because they are grouped in units of their birthplaces. Talismans from the local shrines are carried to the heathen land: a rock from this shrine, a blessed lucky charm from that shrine, a wooden plate from the shrine on the other side from the village. Rumors start: *I saw a soldier I've never seen before telling me to duck, and when I did, a bullet shot through, I never found out who that soldier was, and my family told me that they found a bullet on the enshrined stone back home; we were under attack, and we could've all died, we almost did, if it weren't for the phantom fox that dashed across the snow and drew the bullets toward himself; my mother wrote that the stone statue has been missing from the shrine, ever since the day I left.* The gods are looking after them, men think, and it is true.

June 1940: Shirō, the Letter Writer

One after another, he must write letters to the families of men who have been wounded, who have been killed, and thank god he has to write more letters to the families of wounded than those who have fallen, the martyrs, the heroic war dead, who would have returned to the Yasukuni Shrine to help them in the time of need. It is always the same letter he writes, the letters full of the gallant deeds these men performed in the line of duty, the same words he uses, except that as a commanding officer, he includes his views of how earnest they are, how they served selflessly for the country. Nothing he writes seems real as if he can find no words, the words contained in the bounds of the nation's. He does not write about this man whose thigh bone was shattered by the bullet, and how this man begged and begged not to cut off his leg, but they did, *I'm not married yet, I want my leg*. Or how that man got some sort of sexually transmitted disease, after how many times he told his men to be careful, to think of their wives, their families back home. Another man who shot his own foot off, in the moment of desperation, in the moment of weakness, and they had to cut off his foot. He writes letters as he thinks of the restriction being placed on the home front: sugar for a family of sixteen or less, 360g of sugar. His family. Matches rationed every two months, a family of less than six receiving a small box of matches, and for the family of more than seven, a big box of matches. All these restrictions. Not just his family, but also his men's families. War is encroaching on the motherland, the very land he and his men are fighting to protect.

The Rhetoric for the Family of the War Dead

Bullets rained on (<u>posthumous rank and last name</u>) like hailstorm, yet (<u>posthumous rank and last name</u>) gallantly and fearlessly charged ahead, leading his men with his sword pulled out. As soon as he saw (<u>number</u>) enemies in the trenches ahead of him, he struck them with his sword, butchering them. Others began to scatter, but (<u>posthumous rank and last name</u>) chased after them, killing one enemy after another. Once he was done, he charged ahead, until he was gunned down by machine-gun fire, where the bullet pierced his (<u>the scientific name of a body part</u>). He, like the warrior god he was, charged ahead, even while he was mortally wounded. Knowing he was about to expire, he yelled out, *Long Live the Emperor! Long Live the Emperor! Long Live the Emperor!,* and passed away.

August–October 1940: Shirō Promoted to Staff Sergeant

They spend the month of August marching in rain and mud. They spend that month pushing cars and donkeys out of mud, constantly rained on, knee deep in rain, not being able to cook food, not being able to stay warm or dry. They spend that month fighting blisters and athlete's foot. They spend that month sleeping outside, or when they come to a village, driving the residents out of their houses so they could sleep inside, amidst the chickens and pigs and tired-looking bulls. They spend the time angry, tired, and shooting, especially when the company commander is mortally wounded. Shirō is pulled out of the battle on a mission back to Japan, to pick up new reinforcements because brother-soldiers are wounded and killed, because this is the time when men could still be replaced. And the news arrives to Shirō during a pause in marching: that he has been promoted to Staff Sergeant. He is happy. He is tired. That's all he can think.

October–November 1940: The Homefront, or Behind the Guns

They tell us, no more luxuries like fur, like fountain pens, like gold. They tell us that perfumes are forbidden, that business shirts and men's suits are useless. Stores begin to shut their doors because they have nothing to sell. Why should we live with superfluous things we can live without when our men are sacrificing their lives so that we can live in peace, so that the rest of Asia can live in harmony. No more English words, no more English alphabet. We are Japanese, we must not live with decadent ways of the West. This is the time to embrace what it means to be Japanese, and embrace the Japanese spirit, which is about modesty and austerity. Dance halls close one by one, and we of the Great Empire of Japanese Women's Committee stand vigilant on street corners as we have done with *sennin bari,* but this time to scold, to harass anyone and everyone who still espouses the western decadent way: women with perms, women wearing long-sleeved silk kimonos, women who shop for more food than necessary. We take scissors out and cut off the sleeves, leaving behind shorn-off sleeves. We pull on perm hair and tell them to be ashamed of themselves for wanting to become like western women. We tell men still wearing western-style suits that they must wear the *civilian uniform,* so like the soldier's uniform, khaki, because even if they are not at war, the nation is indeed at war. With every sending off of men, we gather by the train tracks, by the stations and harbors, and serve tea to the mobilized soldiers. We gather for every village-wide *sennin bari* and flag signing, for every neighborhood meeting and making of packages. And a month later, we go from one house to another, collecting metals and jewelry, *for our men, for the war effort,* and we keep track of who is reluctant, who is willing. And the more we collect, the more the neighborhood is praised and awarded. *See, that town, that neighborhood, is united in their war effort. They have given up so much so that our men in the battlefield can win. Look how patriotic they are.* And we of the women of the Great Empire of Japanese Women's Committee, with the symbolic medals and services, are proud: *see, we, too, can serve the nation.* What we do not see is that we are, for the first time, liberated from the duties and obligations of the house in the name of the nation, and we, too, can be part of the rhetoric of the nation.

1940: How the War Dead Come Home after the State Funeral

Even if the box that contains their son or husband is light, too light even for a handful of ashes, loved ones are not to open it, in the same way that no one is supposed to open a coffin once it is sealed. For this box is a coffin, carrying the ashes and the personal belongings of the dead. Their son finally came home, not the way the parents had hoped, walking through the doorway, but as a discarded shell of a god who has already made its way to the Shrine, where all other gods dwell. They do not show grief, though the loss so keenly felt their hearts break as they stand at the village funeral. As villagers pass by with condolences, the family members nod, dry-eyed, thanking them. *Yes, my son died honorably for the country. Yes, he is now a part of Yasukuni Shrine as a god.* They are not supposed to show sadness, for why be sad when their son is now a god, having served the Emperor, the living god in this holy war? Besides, it states here, in the letter from the regimental commander how this son was in a battle, and though he was shot as he charged ahead, he kept going and going, volunteering to be a target so that the others could push the frontline forward, destroying the enemy company. A heroic death. An honorable death. So the family stands at the village funeral, mutely, thanking everyone for coming, the way they have seen others do, the way the mayor and teachers told them was the right reaction to this sudden but honorable ascent to godhood. Only after the funeral, in the darkest hours of the night, when there are no prying ears and eyes of the neighbors, do they sigh in relief, and become immersed in the individual and familial grief that was not allowed to them in the light. Some of the defiant ones might open the boxes — even though they were told not to; sometimes there is a handful of ashes, maybe their personal belongings. Later on, when the State can no longer carry the dead back, they put a fistful of soil where the men have fallen, sometimes a rock, sometimes a name, sometimes the strand of hair they had left with the regimental headquarters before they left the innerland. And now, in the most basic unit of a man's life — in a family — for the first time, family members allow themselves to cry and mourn for the dead who, in their eyes, are not gods, but their loved ones.

December 1, 1940: Susumu Conscripted

A second son has nothing on which to rely. He must take initiatives in order to move ahead. And how proud he was the day he was told he was moving from Matsubara Tax Office, a small branch tax office in Ibaraki, to Kandabashi Office, in the middle of Tokyo. And he was only nineteen years old. He was the chosen one. He printed so many postcards and sent them off to as many people as he could: *Greetings. I am writing my deep gratitude for supporting me, both in public and private life, while I was working at the Matsubara Tax Office. I have been transferred to Kandabashi Office at this time; please continue bestowing me with advice and support at my new work place. July 1937. Kandabashi Tax Office. Shimura Susumu.* How he wrote the addresses on the postcard, announcing his move with pride. He had made it out of a small office. He was now going to be at the center of the whirlwind, a promotional move right on track. Even when the office burned down in 1940 after a fierce lightning strike, he had a purpose. It was not just on this office that his ambition was set, he had the whole of the Ministry of Finance in mind. He has been working in the tax office in Tokyo when the notice comes. His life, he felt, was just starting. Just this month, he finally finished his Bachelor's degree at Chūō University, which kept conscription at bay while his studies lasted, but it kept overshadowing the future. How hard he had worked at it. Working during the day, attending classes at night. He was good with numbers, but he also knew that without university degree, he would not be able to advance as a civil servant. And now that he was done with the study, and as soon as his degree was confirmed, came the conscription notice, then the result. Now that his life is about to start. He is twenty-three years old.

December 16, 1940: New Recruits Arrive

Once again, the fresh, excited faces of the new first-years. Boys. Young men. Shirō scans their faces, and their faces looking so earnest, scared, in awe with their drill sergeant in front of them, and he feels a pang in his chest. He pushes that thought aside, and orders them to run, fully equipped, around the gate.

Slang Dictionary Used by Imperial Army Soldiers in China

— *Pokopen* = useless, only used when talking to Chinese
— *Tenho* = very good
— *Anpontan* = stupid, fool
— *Pi* = prostitutes, usually referring to Chinese, but can be used for comfort women in general
— *Mei farzi* = can't be helped; usually used to refer to Chinese attitude toward life
— *Fairai* = broken, destroyed, usually used with no-ten fairai, or stupid person
— *Heihao* = very very good
— *Coolie* = Chinese laborer
— *Saiko saiko* = to have sex
— *Wanra* = to finish, to die

IMAGINARY DEATH

Fig. 3.9. Kōkichi as a high school student.

— *Yao* = to need
— *Kunyan* = young woman
— *Haohao dei* = good

January 8, 1941: *Duty Is Heavier Than a Mountain; Death Is Lighter Than a Feather*

Soldiers must not surrender. Soldiers must carry the dignity of their nation and the honor of the family in all their deeds and all their actions. Soldiers must treat the enemy prisoners-of-war with dignity. General Tojo issues a new Imperial Rescript in the name of the Emperor. The rules of engagement. This rule, later on, becomes a binding curse as soldiers in the southern islands, hungry and sick, would be forced to march through the pathless jungle, and if they faltered, if they did not show up on the designated time, they were considered as having run away and could be executed on the spot, without trial, by their officers; men with families, who would rather stay alive than to die anonymously in the strange land, alone, who would rather have the shame of surrendering than the pain of staying alive like wild beasts, walking away from their hiding places with the American surrender pass that flew down from the sky like saving grace, could be executed by their desperate officers for high treason; men forced to charge the enemy position with only a stick in their hands as they are mowed down by machine guns; men staying on to defend their position, no matter what, even ten years after the surrender; women jumping from cliffs in Saipan, in Okinawa, because they would rather die than live the living shame of being captured; girls learning how to pull the pins out of hand grenades and embrace the enemy as they blow up; boys carrying bottles of Molotov cocktails to run under the enemy tanks. All this in the name of duty. All this will happen because of the curse placed by their leaders.

January 1941: Susumu at Basic Training

Just as Shirō has feared, Susumu is the target of corporal punishment, both just and unjust. He refuses to speak in the language of the army, he refuses to cower when his drill sergeant screams at his face, his face turning bright red as Susumu looks ahead of him, just as he is taught, his back straight as he is punched, never faltering, never buckling from the impact, from the pain and shock of being punched. *You think you are better than anyone else 'cause you went to university?* a second-year yells at him, the man who couldn't get into the university and whose education was stunted at the higher primary school. *You think you can stay stubborn here?* other second-years sneer, humiliate him for the smallest offense: for refusing to clean dishes, for reading novels when others write in their diaries, for outrunning everyone, even the drill sergeants, because Susumu has always been a good runner and that's what kept him going in the high school soccer team. He stubbornly stands straight as they scream and yell at him, *Mr. Student, let's hear you say that you're a worthless piece of shit,* and he parrots, *I am a worthless piece of shit,* but they see his eyes, and they won't leave him alone, they won't let go of their taunt, until the drill sergeant comes out of his room and calls him in, *Look, I know your brother, Staff Sergeant Shimura, he's a friend of mine. I don't know why you can't be like him,* and Susumu does not say anything. *Look, if you can follow orders, you can be promoted to the Officer Track, and with your degree and background, with good grades and being a competitive soccer player at the national high school level, having worked in the tax office, you'll be made a lieutenant after the officer school. You can go far, and there are enough officers who know your brother, you can have it easy. All you have to do is to change your attitude and accept your situation.* How can he tell this kindly man, *You ruined my life. I went to the university at night and got my degree while working at the best tax office in the country. My life was starting, and now, I'm imprisoned here in this hellish basic training. All I want to do is to get back to my life that just started. All I want is to go back to my old life.* But instead, he keeps his face impassive. The staff sergeant sighs, and shakes his head. Susumu does not know that this moment sealed his fate, and that he will remain a second private, and his life does not amount to much as a private.

March 1941: Susumu's Last Photo

Susumu stands in front of a bush in this photo. His expression is blank—or perhaps a grimace or perhaps a scowl. He stares straight into the camera with his back straight. He is twenty-three years old, a conscripted man, with only one star on his lapel. He does not want to be there; he never wanted to go to war. But here he is, forced to have his photograph taken, to be used as the photo to adorn the altar if and when he dies. He must have written his last will, as was the ritual for all conscripted men during war. His older brother, Shirō, is in China, and rumor has it that he will be discharged from the army after nearly four years, two years as the Emperor's Guard, and two years as staff sergeant, in charge of training new grunts, in China. Now that his life is almost beginning to start. Why now, he must have asked himself. He had escaped from the farm, from the land that could have bound him if he were the first son. Now, he is a private, not a civil servant in the tax office. He is one of 630,000 new drafts this year. He is now Private Susumu Shimura in the Field Artillery 14th Regiment Company 7.

March 15–19, 1941: Shirō Leading the First-Years in Battle

The finishing of training: taking his men into the battle, their first battle, and he slaps, he yells, he screams at his men to stay in position, to keep shooting, to move forward, and they are scared shitless, as all men are in their first battle, and he punches them to make them move. *Go, go,* he yells and men crawl forward, slowly, almost like babies learning to walk for the first time, and *faster, faster,* and they try but they trip over their rifles on their back and their backpacks get in the way, *second platoon, cover them, shoot, shoot,* and men shoot with their eyes closed, their bodies jerking on impact, and their helmets falling over their eyes, and *shoot, goddamn it, keep shooting,* and the first man in the first platoon reaches cover, the bullet-riddled cottage, *hand grenade, throw the fucking hand grenade,* Shirō screams but his yell is erased amidst the storm of gun shots, but the man remembers, he is a good soldier, a fourth-year who knows what he's doing, he pulls the pin and throws the hand grenade from the open window and crouches down and the house hiccups loudly, rising dust from the windows, but the bullets keep coming from the invisible direction, *cease fire, cease fire,* Shirō screams, and all his men stop firing, but the enemy keep firing, *two o'clock, shoot, shoot,* and the air turns thick and hand grenades fly through the air, and Shirō thinks absentmindedly, *need to remind them to conserve the grenades,* and just as he thinks that, a Chinese stands up with his arms high, and he is shot down, riddled with bullets, and another tries to make a run for it, but he is shot down as well, the first-years can't stop the battle fever that kept them going, they can't stop their fingers from pressing the triggers, and another runs through the tall grass and the bullets get him as well, one by one, the enemy is shot down. For a first battle, a good and easy fight. A victorious fight to raise the morale of the first-years.

1941: Monday, Monday, Tuesday, Wednesday, Thursday, Friday, Friday

All of us are involved in the Holy War, the sacred war. Every day of the week is a working day. The voice over the radio dictated by Central Command tells us the whereabouts of our men as we put red pins on the map of the world, tracing the path of victory that is surely ours. Central Command tells us so. The world outside is changing: men walk by clothed in the color of the military. Men too old to serve in the military, men who have yet to be called to serve, men who have served already, all dress like soldiers, their heads shaven close to the skull, wearing puttees, just like the Germans did in the World War I. They can go to war anytime, but then, we are at war, all of us. The rations are in; English words are slowly disappearing from restaurants and cafes, no journalist can speak English at the news press conferences, rice is rationed. A person between eleven to sixty years old: 330 g per day. Flour, rationed: for a family of four to seven, 565.5 g per month. Fish: 112.5 g of fish per person a day. Cooking oil: 541 ml for every two-to-three-person household every three months; and dog and seal meat as well as seahorses can be sold at butchers. We are now at war like we have never been.

May 1941: One Battle after Another

A command: Shirō is to be discharged honorably for serving the country for nearly four years. A command: to lead his men into battle, one after another, on a long mission. *Man, be careful,* other sergeants tease him, *don't want to jinx you, but men about to be discharged have the higher rate of getting killed, just be careful, don't relax, okay, don't get yourself killed,* they tease him, but they mean it. They've seen enough men getting killed a month, a week, even three days before they can go home, as if the moment they find out they are going home, they let go of the battle instinct, and though they're not taking risks, the tension that kept them alive so far is gone. The animal instinct. The survival instinct honed after so many days and months in the battlefront. Shirō nods and goes off, leading his men. But all he can think of is things to buy to take home as presents, and Masa as big as a moon, almost due from what she wrote in the letters, and he wonders if she is getting enough nutrition, with a meat ban in effect. Would she eat dog meat that is said to be sold now? But he can't relax, he tells himself. He needs to go home. He needs to stay alive for one more month. And as a staff sergeant, he can't let his men die or be wounded because of his inattentiveness. Yet, all he can think of is what he will do when he gets home. That's all he can do. His mind has been released, and even before his physical body is making its way toward home, his mind is reaching out, touching, living the everyday life of his home.

May 13, 1941: The Birth of Yoshiko

A twitch. Then the familiar sharp pain. That's what Masa feels as she bends over the rice paddy, because it is that season, because every May for so many generations, this is what rice farmers do — past, present, and in the future. And all around her, ghost hands hold the baby stalks, pushing the roots in, then twisting their wrists so that the roots take hold, so that they are mud-bound, so that during the rainy season, they hold fast until late August, when they'll turn golden and be bountiful for the harvest season. But that's many months from now, and now Masa stands knee-deep, her stomach as big as a moon, making sure that the stalks are where they should be. And that familiar twinge, just like the ones she had when she had her first baby five years ago. The baby is coming. She waddles awkwardly, lifting her heavy leg, one at a time, and climbs up the mud dyke on her knees and hands. Her father-in-law looks up questioningly, and she tells him the baby is coming. He nods. He knows that there is no place for a man in the birthing process, and goes back to touching the stalks, one by one. She holds her stomach as she waddles back to the house she can see far away in front of her. She can feel it, an easy birth, the way it was an easy pregnancy — that's what she thinks as she reaches the empty house; everyone is in the field or at school, and she is the only one home. She gathers the wood and starts heating the bath water. The sharp pain. How this is so different from the previous birth, when she didn't know what was going on. She knows that birthing can be easy, but it can also be difficult. The water is ready.... She washes every inch of her body and once done, submerges her heavy body into the water and sighs. *I wonder,* she thinks, *I wonder if it's going to be a boy again.* And after another sharp pain, she shakes her head. *No, a boy-child would stick out more, that's what they say about boy babies. Maybe it will be a girl.* The bath. It's done. She is clean, and she is ready to settle into the birthing room and wait for the midwife to arrive. And suddenly, she doubles over from pain like no other. And suddenly, she can feel the child starting to slide downward, downward. Masa kneels down and the child comes slipping out, out, out. A girl. A girl crying between her legs, an umbilical cord still pulsing, pulsing. And that's when the midwife enters, smiling, and the child cries happily, and there's a new child: Yoshiko, the beloved child.

June 11, 1941: Leaving Heze

The night before, a feast to wish him goodbye and good luck, a feast to wish a good life as a civilian. The gramophone playing jazz, and bottle after bottle of alcohol. It is early in the morning still, yet his men have come out to bid him farewell; his superiors come out to bid him godspeed. They salute as he gets in back of the truck, and some of his men in near-tears, *Staff Sergeant Shimura, we will keep fighting for you,* and Shirō is touched by their childlikeness, by how much they like him after all the yelling and screaming he did, and he, too, will start crying, if he lets that emotion get hold of him. So instead, he laughs and says, *You make it sound like I'm dead, but I'm alive,* and they all laugh and cry at the same time. As the truck revs, his men clamor around the truck, following the truck as it slowly moves away from the garrison toward the town center, and men keep running after the truck and keep waving at him, waving and running until the truck becomes too fast to catch up with and they stop and keep waving, and Shirō yells out, *You need to run faster!* The truck runs through the town he did not care about. It is the same. The Chinese vendors yelling at each other. Men and women moving around busily. The prostitutes lounging like cats on doorsteps. The sizzling smell of sesame oil and chili peppers. The full textured smell of roasting meat. Dogs scattering as the truck rolls past them. Then through the northern gate where he spent so many hours and weeks training his men, through the dusty dry road that cuts through the wheat field that goes on and on. He looks outside to remember all this, to memorize this, he keeps his eyes on the landscape so he can tell others, if asked, that the sand got in his eyes and that is why he is crying.

June 14, 1941: Arriving Home

He chooses not to tell his family when he is arriving. He wants to surprise them, to see them not in their finest, standing in attention at the station, but caught unaware in their everyday gestures and habitation. His backpack is filled to the lid with presents he bought: jade stones for his wife, scrolls, cigarettes, full-color photo books of China for Seijurō, and many more that he bought while in Qingdao. With each step he takes, he feels lighter and lighter, as if the weight he has been carrying on his back is flying away like birds. The very path he took to school every day, he is retracing it. The green rice field. The river. The straight path along the rice paddies. Houses dotting, emerging out of the green. With each step under this brilliant blue sky, he is coming closer and closer to home. And he is home but he is not the same man who left home four years ago.

June 25, 1941: The Hero's Welcome

This is his twenty-fifth birthday; though he is not superstitious, he can almost believe that the gods have their hand in this. He arrived in Ujina Harbor, near Hiroshima only the day before, from Qingdao on a ship, only a four-day trip. How close China is from his home, but so far away; not far enough to forget all the things he has seen in the last two years, and close enough that he cannot stop looking toward the land, toward home, toward his future. And in four days, he will step into his home by the inlet, where the river tastes of the ocean and herons glide through air, seagulls diagonally cutting the sky. So much to do before getting on the train: shaving, dusting off his best uniform and slipping into it, shaking off the images he has seen on the battlefield, buying a gift or two for his children, a boy and a girl, but most of all, keeping his spirit from leaving the body, to see Masa. He is going home, as a hero, as a released soldier, to his wife.

Fig. 4.1. Shirō in 1941.

IV

Domestic Life

June 1941: The Drowned House

It rained so much, because this is the month of water, because the gods are providing for the harvest to come, and water is necessary. But who would have thought that it would rain so much, day after day, night after day, slapping the earth, providing more to the river's supply until the river god reared his head and flooded the entire riverbank, swallowed up houses by the river, rice paddies, washing away the young rice plants, carrying people off. The house by the river is under water: on the little boat, Shirō carries Seijurō on his back and Masa carries Yoshiko and the letters Shirō sent from China as they row through the house. Shirō doesn't know that Masa is carrying his letters on her back. He doesn't know that the bundle of letters will survive sixty years, only to be discovered in the corner of the damp storage, in a torn-up box, in pristine condition, the last pages on top, exactly the way Masa read them, reading the last page and folding the letters back into the envelopes. For now, they are safe: their entire family. Yoshiko will later tell this story again and again: when I was born, that tiny river flooded and we had to escape in a boat.

July 1941: The Restlessness of the Dream World

He is pinned down on the field; bullets fly all around him, but he cannot decipher the sounds anymore. He looks around to see whether his new recruits are doing what they are supposed to be doing, this is their first battle, but no one is around. Bullets stop. A hand, then an arm sheathed in the familiar dry-earth yellow, but not the body. No blood. Then he sees. A leg still encased in a garter, tight, but no thigh. Limbs scattered through the field. The earth is quiet, and the sky is so big above him. He calls out the names, *Yamada, Tanaka,* but no one answers. A dog bays; it is night, and no one is here. He has been left behind, and he knows that anything can happen to a Japanese Imperial soldier left by himself in this land. No language. No kindness. Only hostility all around. He must get back to the base as quickly as he can, but the stars above are not the stars he can read and the pack on his back is so heavy that he cannot take a step. He must get back to the base, he is vulnerable here, out in the open, who knows how many of the enemy are hiding in the tall grass, but the pack is getting heavier and heavier. He pulls his pack off, and opens it, only to find decapitated heads, heads of his recruits, Yamada, Tanaka, Nagai. Shiina, his best friend. He wants to scream but he knows that he can't, so instead, he claps his hand over his mouth but he feels he can't constrain it, he feels that the scream will escape him, his body, the body betraying him, the wail that's about to come out, pumping his stomach, expanding his lungs... and he wakes up, and Masa is sleeping next to him, and in between, Seijurō, their son.

July 15, 1941: Susumu Comes Home from Tokyo

The order: he, along with his company, will be shipped out at the end of the month. Susumu has packed up his apartment in Tokyo, where he has lived for the past four years, and is home on leave, just before his journey north, he hasn't been told where yet. He has gotten a college diploma while working during the day. When Shirō told him to take care of himself more, Susumu had laughed and said that this was for himself, that he would not have it any other way. He was on his way to working in the Imperial city, appropriate for a landless second son who will not get any inheritance, but given a chance away from the land, where men and women age so rapidly. Shirō and Susumu sit together by the river with a bottle of sake. This is the first time both of them are wearing civilian outfits: for the past five years, one of them had belonged to the State, and when on leave, they were expected to be garbed in the uniform of the Emperor's children. They sit, not as brothers, but as two best friends saying goodbye, as they sit by the river in which they swam so many hours together, the current intimate to them, every bend as familiar as each other's unuttered thoughts. Susumu has lost so much weight, and he carries anger with him like a skin, as if this is the new self, and the boy he knew was no longer here. But then, Shirō himself has changed, he knows that. Seijurō, his son, is shy around him; he loves Masa, more so now that he has come home alive, but something is amiss. Maybe it's the intimacy of letters, where he could trace his thoughts, and present a better man, a better self, an intimate self, but now that he is home and there is no need to write letters, he does not quite know how to tell her what he used to write to her, *our hearts are one.* And he is not the same around Susumu either. But after all that he has experienced, he's not sure if two hearts can be one anymore. He does not look at Susumu; Susumu does not look at him. They both face the river. *Will Susumu survive the war,* he wonders, *with his rebellious nature, with his unwillingness to submit to anyone, his stubbornness? He will have a hard time, he already did in the basic training. With his education, with his athletic prowess, he should have made it to the officer's school, but he didn't; what is to become of him on the battlefield?* As if he heard Shirō's words, Susumu says, *I'll be working in the accounting department of my regiment, and you know how field artillery rarely see combat.* Yes, Shirō thinks, they rarely have casualties. Yes, Shirō tells himself, Susumu will be alright, he may still have a chance to come back alive like Shirō.

Fig. 4.2. Private Susumu Shimura.

July 26, 1941: ABCD

It's no longer the Soviet Union that is a threat to the State, but ABCD, *Americans, British, Chinese,* and *Dutch*. And today, America and Britain have frozen our assets, they say that they will no longer trade with us. They were not supposed to be against us: they were supposed to be observers, they were supposed to be understanding, no matter how much they might have criticized our actions in China, in Manchuria — after all, they're doing the same thing all over Asia. Who are they to tell us what we can and we can't do? But now that we can't export any more silk, now that we can't import any more oil, how are we to carry on our holy war?

July 29, 1941: Private Susumu Shimura Entering Field Artillery Regiment 14 Company 7

This is as far as Shirō will take Susumu, to Sakura Station, where the train will take him to Utsunomiya, then to an undisclosed port somewhere, and Susumu will get on a ship with the rest of his company up to the continent. The village ceremony went as usual: Susumu stood in front of everyone and gave a speech about how he would serve the village and the country to the best of his ability, the formula of a speech, and with the expected three cheers, *Long Live the Emperor! Long Live the Emperor! Long Live the Emperor!* Someone yelled out, *Kill as many Chinks as you can,* and Shirō was startled to find that it was an old man from a couple of farms away, a man known for his mild manner. Shirō knew that Susumu did not mean a word of his gallant speech; though a part of him, the *veteran* part of him, believed that this war is holy, just, but while there was something to be said for how being a soldier taught him much about life, a part of him also knew that it was not at all that people make it out to be. There was no such thing as the heroic death, or perhaps, because they died in battle, all deaths were heroic. Men boasted about how many *Chinks* they killed that day, and even though each and every one of them were scared, so scared that they literally shat in their pants, they would not say it. They say that after the first kill, it gets easier, but it never does. How many times did he shoot, making sure that he would miss the mark, only to shoot right on target so that he would not be bothered by their screams, by their suffering? That there is so much more, so much to being a soldier. But nevertheless, Susumu needs to be a man, because being a soldier is what defines a man in this time, and without it, he will forever be laughed at, ostracized, never becoming a real man in others' eyes. They have walked this far together: Shirō wearing his veteran's uniform, as the times demand, because even as a noncommissioned officer, his body still belongs to the State and the Emperor-god until he is forty. He must be ready at all times to be called to duty, if the Emperor-god demands. Susumu wears an already tired looking uniform, his back rounded from the weight of the uniform, though Shirō knows because his brother is his best friend, that his brother does not care about honor and service in the way he does. If it were another time and another place, Shirō would have offered to carry his burden, but it is this time, this place, and Shirō knows that men must carry their own burdens, and if they cannot, they will be left behind. The station is in front of them; people crowd the small platform, seeing their men off. The train pulls its whistle: it is time to go. Susumu and Shirō look at each other, waiting for the other to say something, anything, but the words are stuck. The only thing Shirō can offer him is g*anbatte koi* — do your best and come back. Susumu climbs the steps into the train, and Shirō follows him with his eyes as far as he can, to the left, the right, he finds him on the car to the left, follows him walking, keeping their eye contact, walking to the middle of the car to find an empty seat. As the train pulls away, the cheers, *long live, long live, long live,* but Shirō does not say *the Emperor*. Shirō cheers, *Banzai, banzai, banzai — may you live a million years.* They do not know that this is the last time they will see each other alive.

The Farm

He has been home for nearly a month now, and he has returned to the farming that he has dreamt about for four years. All the things he had learned at the Agricultural College in another life time come back to him, as if he never left this land, his land, this farm four years ago. This is his home, and he understands the land as he never did, because he loves it, because he dreamt about it for such a long time. When he was on the continent, it was all he ever dreamt about, tasting the soil, reading the sky, working side by side on the land he so loves. Which is better? Holding a rifle, pulling the trigger, aiming at the *Chinks*, and falling asleep, trying to tell himself that despite how futile it may be, he is doing this for the country, for the Emperor, for his family back home? Or bending his body in half, kneeling on the soil, worrying about whether the typhoon will come, whether the harvest will be bountiful, readying for winter, and falling to bed to the nightmares that wait for him on the other side of sleep? He is made restless by the repetitiousness of the everyday life, the ordinariness, where he is who he is, and nothing more. He should be grateful, he tells himself, there are so many who didn't come back alive. He should be grateful, but there is something amiss, something he cannot quite name.

August 22, 1941: Susumu Arriving on the Continent

This was the first time he was on a vessel on the ocean. This was his first time traveling outside of Japan. The boat ride was stuffy, tight, they were housed with horses and slept in bunks spaced in threes from the floor to the ceiling, giving him only 80 cm of privacy. He could have written a letter home, but what is there to write about except that he is on a ship, crossing the ocean, and all the things he wanted to write about would have been censored anyway: that they have given him a winter coat, though it is still the middle of summer and the heat is sweltering, so he thinks that they are going up to the cold northern land, where Manchuria is; that he has landed on Juhua Dao, named Chrysanthemum Island because it looks like a flower blossom, a fitting place for the Imperial soldiers to land; that he has walked around on the island, and how he likes its wildness, and just across in the distance, he can see a city. All of that would have brought him a slap from his officer, no, a punch, telling him that he needs to follow the postal code, he cannot write about where he is, where he is going, because those are state secrets. He has been on a train for two days, on a cattle car; he has slept in uniform with a loaded gun, and how he stinks. Yes, how he stinks now. They have been told that this is a green zone, that the rebels have been suppressed, that it is an easy journey to where they are going, though no one has told them where they are going. Up north, someone whispers, and that becomes truth, then someone says outer Mongolia, and that becomes another truth, and the truths collide, wrestle, struggle to take hold, until another utterance is passed from one mouth to another, molding itself into truth. But he is excited. He can see from the tiny slots between the boards the vast landscape, a green field that keeps going on and on, such a flat land, and he sees that the land does not end, not like it does back home, but keeps going and going. How big the world is, he thinks. He never knew that the world could be so big, and there is so much to see, so much to live for.

October 1941: Kōkichi Leaves for Basic Training

Kōkichi is not wearing his finest to go into the basic training, as Shirō had done when he first entered that barrack so many years ago. Kōkichi is careless, he always was, and he doesn't take anything seriously. That's what worries Shirō as they walk together toward the station: how will he survive basic training, where men are punished for no reason, where men are broken into pieces and rebuilt, as he had done to his own men, making their minds submit to his command without questioning, without thinking. But then, Shirō thinks, Kōkichi may be the one who, with his charm, with his laughter, will make it through, because he has never been that earnest, because he has done everything lightheartedly, but never has submitted to anyone. Kōkichi looks at Shirō slyly, *Are you worried about me, big brother?* Shirō can't help but to laugh at that question, *No, there's no use worrying about you, you always managed to squeak by at the last minute.* He has already told Kōkichi what he needed to tell him, *Get in the officer's track, with your education and intelligence, there shouldn't be any problem,* and when Kōkichi protested that he didn't want to be an officer, he just wanted to do two years of duty and get out, Shirō just said, *Do as I tell you* and Kōkichi submitted because Shirō was an older brother, because he was the next patriarch in line, because he respected Shirō, always saying, *It's hard to be the youngest brother of such two great men, so intelligent, so successful in school, doing everything well as they do.* Shirō did not say, *When you get to the officer's track, they treat you better — you get better food, better housing, your life is worth something in the eye of the government, you have a better chance of staying alive.* As they reached the destined meeting place, Kōkichi tells Shirō to take care of himself, to look after the family, and Shirō nods, *Do your best to serve the country.* Kōkichi walks away, and suddenly, he turns around and flashes that mischievous grin of his, *See you later,* as if he was just leaving for a walk, a short absence, and he will be back soon.

October 10, 1941: Susumu Arriving in Shanghai

They never told him where they were sending him. Two months in Panjin, where winter was threatening to arrive, training for combat in winter though it's still autumn, where he learned about how flat the land was, and when, as they stood in the field looking at the horizon — they just could not keep their eyes away from that line in the distance — one of his friends, in a sentimental moment, said, *On the other side of the horizon is Europe,* how true it sounded. And now, he is on a boat once again, and is this time issued summer clothing, and he knows he is going somewhere south, though he's not sure where. Someone whispers that they are going to the South Sea, and someone starts to dance, thrusting his hips, an imagined dance of the natives, and they roar in laughter. Then someone says that they are going to some small island where there's only *dojin*, earth people, and they will be eating bananas until they get sick. In the past months, Susumu has learned to like these men, the men in his company, his unit; they are like any other men, some he may even have met at college. Here, a man who used to be an owner of a sake distillery, who can tell you everything there is to know about rice, water, and drinking, though he never drank much; there, next to him, is a college graduate like Susumu, someone who took classes at night, just like him, and they got along well. There, the joker who makes everyone laugh at tense moments, whom everyone liked. They compare notes of home, they talk of their families, their wives, they tell tall tales, and they talk of missing their homes. Just like he does. They are not too bad, he thinks. Now, they are heading south, to an unknown destination, his life is no longer his own but with this unit.

November 1941: Shirō, Mr. Soldier, the veteran

Everywhere he goes, people call him Mr. Soldier, they tell him that he made them proud, but he carries their words uneasily, as if he does not deserve it, and he doesn't. Only he knows that. Whenever he heard that a new recruit who trained under him had been wounded, or even worse, died, always he thought, *Was I not good enough as an instructor? Didn't they remember? Did I not instill enough combat instinct in them?* He has tried to cut down from three packs a day to one. He has tried to tame the nightmares. He has tried to say that nothing is wrong, but there is something wrong. He has finally gotten used to the idea that there is no need to sleep with a loaded pistol under his pillow, and instead, he has gotten a rifle, but the rifle remains within an arm's distance, just in case. Just in case of what, he wonders. And when things get too much for him at night, he can quietly slip away from the bed, take a pack of cigarettes, and walk up and down the river bank, or, on a clear night, when the moon is out, he can take the boat and ferry upstream, up, up, until his mind is cleared and he can go home to where his wife and two children lie sleeping, and slip back into the warm bed and be safe again.

November 1941: Holding Seijurō

Simple joy: Seijurō running up to him, calling him *papa*; Seijurō following him around with his short legs, asking simple question, *What's that?*, not like the way he was when he first came home and Seijurō hid behind Masa, *Your daddy's home, go to daddy*, and him hiding, hiding, shy around a strange man who was gone the first four years. And today, his little daughter, Yoshiko, looks at him solemnly, and then laughs. Waking up to the sound of a house that is already half awake; the rhythmic sound of the vegetables being cut in the kitchen; the faint smell of miso soup; persimmons gathered, then strung together and dried in the shade; pigs snorting in the yard, and hens clucking; a boat slowly making its way downstream, the pole knocking against the side of the boat; the faint smell of the ocean carried far inland, to where he is; a white heron gliding through the sky, then tilting its lower body underneath, the wings opened wider for the slow touchdown onto the river inlet; sweet potatoes cut into slices, then left to dry out in the sun; Masa touching his shoulder, *Wake up, it's time to get up*; the morning defined by the cold, made tangible by the breath. All of this. How he loves it so.

Wintering Land

The winter has come, even to this farm by the river. No need to work on the land, but there is much to do to prepare for next spring's planting, and to live off of what they have harvested. During autumn, the cold wind from the ocean blew through the hamlet, tearing off leaves one gust at a time, and now, this is winter. Inside the house, so many bodies: his parents, his wife and two children, his siblings. The house is kept warm by everyday conversation, by the ordinary gestures of his family, by the idea of home, by worries for Kōkichi in basic training, and Susumu somewhere in southern China, by the thought of teaching again, this time at the Young Patriot's School, young farmers who couldn't go beyond elementary school because where they live, it is not intellect but strength that matters, though he knows better. He will teach them that education does matter, that it can take you far, perhaps away, if they so wanted. But he understands the rule of this village: first sons are bound to the land. It is a comforting thought. All is good, he thinks.

Unknown Time, December 6, 1941: Strange Message on the Short-Wave Radio

Eastern wind clear sky, eastern wind clear sky, the message keeps repeating itself on a certain channel on the short-wave radio. It has no meaning, is not meant to have any meaning, except for a select set of ears. And when it reaches them, all over America and Europe, consulate and embassy workers get up and go into the hidden cabinets, take out the most top-secret documents and begin to burn them, erasing coded messages and documents. They are getting ready.

December 7, 1941: *The Japanese Government Regrets to Have to Hereby Notify the American Government That in View of the Attitude of the American Government It Cannot but Consider That It Is Impossible to Reach an Agreement through Further Negotiations*

The Japanese representatives are twenty minutes late to meet Secretary of State Hull. The meeting starts late. They greet each other, as they are diplomatically trained to do, but something is wrong. Without saying much, the representatives hand the

Secretary a document. It is quiet. Only typing sounds outside. But the air within the office tenses up; it cracks. The Secretary's face, from mild curiosity, turns doubtful, indignant, then red with anger while the Japanese ambassador sits impassive, his emotion veiled behind the inscrutable *Asiatic* face he has been trained to put on. "I must say that in all my conversations with you during the last nine months I have never uttered one word of untruth. This is borne out absolutely by the record. In all my fifty years of public service I have never seen a document that was more crowded with infamous falsehoods and distortions — infamous falsehoods and distortions on a scale so huge that I never imagined until today that any government on this planet was capable of uttering them," the Secretary shouts. Two sides facing each other. No longer individuals but two nations at war, though war was not declared in the document. At 7:48 a.m. in Hawaiʻi, the Imperial Army attacked Pearl Harbor from the sky above. That, in itself, was the declaration of war.

December 7, 1941: The Special Force's Attack on Pearl Harbor, Hawaiʻi

Amidst the sunken ships, smoke, all of the easy attack, an easy war that started out with the codename *Operation Z*, the Imperial planes carrying the sun on their wings roared above, overjoyed by how it all turned out. So many ships sinking helplessly like *bonitos* during the high fishing season, their sides glistening with silver scales. So helpless were these soldiers, so lazy on a Sunday morning, running around like ants. The pilots are happy; they didn't know, until this day, that there are no dogfights in the air, that this war will be done before they know it. *Proves the American decadence, we showed them,* the pilots sneer. On the ground throughout Hawaiʻi, forty-nine civilians lie dead, thirty of their last names so much like their neighbors' back home: *Oda, Hatate, Hirasaki, Soma, Kimura, Izumi, Ohta, Koba, Tokusato, Nagamine, Kondo, Harada, Arakaki.* A two-year-old girl with two names, one of her new home and one of her old home: Shirley Kinue.

7:00 a.m., December 8, 1841: *News Flash! News Flash! A Special News Bulletin Released by the Central Headquarters Army and Navy Division at 6 a.m. In the Early Hours of December 8, Our Imperial Army and Navy Declared War on the American and British Armed Forces in the Western Pacific Ocean!*

Ten news flashes on this day, military marching songs crowding the houses through the radio. People are happy. *We showed those Americans!* Children yell out on the streets; men and women come out, cheering and laughing, holding various sizes of the flag of the rising sun. In his village, Shirō does not know what to say. He knows that his students will be taken, one by one; his two brothers have already been taken. And suddenly, between the words of the excited voice on the radio, he sees an image, so vividly, as if in a prophecy: he, too, will be taken.

December 8, 1941: A Myth — Birth of the Nine Warrior Gods

… they spent a dozen hours under water until the hour when the moon came out. With the moonlight, they found an enemy ship left in the bay, an Arizona-type battleship. As soon as they found it, one special midget submarine shot through the water like a seal. In a second, a big explosion sounded inside Pearl Harbor with steel fragments shooting like fireworks, and when the light subsided, no battleships were left on the water. The time: December 8, 16:31. Two minutes after the moonrise. On the same day at 18:11, a communication from the special midget submarine: Successful Attack. *Oh, they have succeeded in their mission. What went through the hearts of these brave warriors? The mother ship eagerly awaited their return, but in vain. No matter how long the ship waited, the warriors did not return. With the last communication at 19:14,*

these five special midget submarines never came back. Oh, these selfless, loyal special midget submarines!, an article boasts. The country is drunk with the victory of Japan against the Western power, victory over America, and how easy it was. The brave pilots shot out of mother ships like eagles from the nest and destroyed, one by one, the enemy ships. How complacent those Americans were, bloated in false pride, so lazy on Sunday! Look how they scattered like ants. And nine of our brave soldiers in the special midget submarines slammed against ships, blown to smithereens. They gave up their lives, willingly gave up and sacrificed for our country, for our Emperor, for us, we cheered. (What happened to the tenth one, the one who appeared in the group photo of the special midget submarine unit? He is the first prisoner of war, moving from one prison on the American mainland to another throughout the years.) We go out in the street and cheer and praise the Emperor. (The one who is captured alive will be erased from history; his family will be ostracized from their community; they will live in shame as long as the war lasts.) Our brave soldiers, they showed how we are to live our lives! The nine warrior gods (but do not talk about the tenth one)!

December 8, 1941: The Unknown Weather

From this day on, this is the land without a weather forecast. No one will know, for the next four years, what the weather is going to be that day except for the Emperor–god, who will receive a phone call every morning, *Your Majesty, it will be sunny today. Your Majesty, it will rain today.* The enemy may be listening in, Central Headquarters reasons, and we cannot give vital information such as this to the enemy. But this is nothing new to farmers, who can detect the temperature, the weather, from their experience with the land. Shirō has been away from the land too long. He is still relearning the landscape of his home. He is still trying to remember the language of the earth.

December 12, 1941: The War

There is war fought in the European theatre and northern Africa; there is war fought in China. Now, these two wars unite, based on the treaty between us, Germany, and Italy, and they, too, declare war on America and Britain. Now, today, the Central Headquarters renames the war from *Sina-jihen,* the China Incident to *Daitoa Sensō,* the Greater Asia War. Farms and houses, deserts and forests, islands and the sky explode with bullets, with fragmented parts of men and women, and no one, none of the people involved in this war, will be the same.

December 24, 1941: Fall of Hong Kong

The excited voice on the radio screams, *Hong Kong is now ours, our gallant and brave soldiers of the Imperial Army have taken over the island of Hong Kong which is under the British rule....* Shirō listens carefully but that is all. He scans the newspaper to see which unit, which regiment is involved in this offense. He knows that Susumu is somewhere in Shanghai, and Shanghai, though still relatively stable, is like an arsenal, ready to explode, and because of the way the army works, they can say that units are in certain areas when they are really somewhere else. Field Artillery. Supporting the infantry. Susumu working in the rear. But still.... In battles, no one remains in the rear; at the end, when things get rough, everyone becomes a combatant, that much he knows, as he remembers the men in the field artillery when he was in China, their mortars gone, had to join the infantry though they did not know the first thing about close combat. Shirō takes a drag from a cigarette and frowns, worrying about Susumu somewhere in China. He could be anywhere in China.

1942: Molding of Little Patriots

Teachers stand in front of blackboards all over Japan, screaming, yelling the slogans of the nation. Children sit straight up, earnestly, remembering each and every word that comes out of the teachers like a bullet. Words are weapons. Children are not innocent. They know that they will be rewarded for being good children, rewarded with praise, with attention, which will raise their status in the eyes of other children. They listen carefully to the rhetoric: *we will not want until we win; we will not stop shooting even when we die.* They carefully listen to the radio and put red pins in places where battles have been fought and won. They chew their food as many times as the Emperor–god's age, forty-one times, before they swallow, just like their teachers tell them to. They are earnest because this is how they survive in the classroom. Boys dream of becoming soldiers, and girls nurses as they sit at their desks, listening to teachers talking of Saipan, Guam, China, America. Teachers tell them that their lives aren't theirs, that they on loan from the Emperor–god. *Your life amounts to 151.1 cm.* A life the average height of a man. They teach the students, *if His Majesty the Emperor wants it back, you must give it back freely,* and the students take note. They learn how to stab straw figures with bamboo sticks; they take their shirts off, even in winter, happy to share the burden of the Imperial soldiers. They ignore a girl who tries to cover her budding breasts; she will be punished, and she is, with a slap from the teacher. On the auspicious day, when the portraits of the Emperor–god and the Mother–Empress are taken out from their shrines, the children bow deeply, not daring to raise their eyes for a glimpse of the sacred. They look at the swirling patterns of the wooden surface of the desks as teachers single out a student whose parents have been arrested: *You are just like your parents, red, red, red.* And later, that student will be picked on, bullied, because children know the hierarchy, and the weak must be punished. Teachers rule the world, and students are obedient. Without knowing, they tell themselves, each in each, *We have seen what can happen to those who rebel, we must not stand out.*

January 10, 1942: Private Kōkichi Shimura, 102nd Infantry Regiment 8th Company

He has come back on leave for several days, as is the custom of all men shipped off to battlefronts all over Asia. Kōkichi made it through basic training, *a breeze,* he laughs as the entire village comes out to say farewell and *Long Live the Emperor* to him. Shirō believes him as he says that, *a breeze,* like the way how he suddenly decided that he was applying to a Teacher's College after attending the agricultural school, and Shirō and everyone had thought it foolish, *Don't waste your time, no one gets into the teacher's college after agricultural school, you are already off the track from what they want...* and after three days of cramming, he took the exam and passed it with a high mark. Kōkichi was like that, bright, intelligent, but never took himself seriously enough. And even while he was teaching at the girl's high school, he would cram for the day's lesson ten minutes before the class. He would do anything to be as carefree as Kōkichi, Shirō thinks. But can he ever remain the way he was before...before that time in China? Can he ever write poems and write in letters, *our hearts are one?* And can Kōkichi remain unscathed by the war? No one came out of the war unscathed. All men were maimed, whether visibly or invisibly, and the invisible scars were more toxic, more festering, because the eyes could not see it, because the others could not see how deep the scar burrowed into the core, and it is the scar that would not go away, and it will never go away. And only when he is with his own men can he feel comfortable, because they do not have to talk about it, because they understand without words. Shirō thinks to himself, *Let my brother remain himself, let him not see what I saw, let him come back home alive,* again and again until it becomes the prayer in the middle of the night, his lips mouthing the words again and again.

January–March, 1942: Cities Falling, Victory Is Just Around the Corner

The voice on the radio, the conduit of the prophets of the Emperor-god, screams out ecstatically, *Kuala Lumpar is now in the hands of the Great Japanese Empire; Singapore fell easily; Java; Rangoon.* Cities in Southeast Asia, under the marching of the Emperor's soldiers, fall one by one, cities that used to be ruled by the Western imperialists, submitting easily, not ashamed of surrendering. The Imperial soldiers march on, if not with the war of the excess materialism, but with the will, with *yamato spirit,* the Japanese spirit, the belief of the sacredness of their mission, *We are fighting the war to free Asia from the Western aggression, if we do not fight this war, Japan, too, will fall like the rest of Asia, becoming the dogs of the western countries.* We go out on the streets with the news of a city falling into our hands, we march as the soldiers do, in line, with lanterns. We celebrate. We are drunk on the victory won by our men, though we do not know that victory is won by men and women dying on streets, on battlefields, with blood making the earth uninhabitable.

1942: China and Everywhere Else

No one really wants to pay for a cheap fuck. Why would you when two-thirds of your salary was forcedly put into the savings account? You still have to pay for your booze and snacks, if there's a PX nearby; if you left behind a family, you'd want to send home some money, too. As it is, your wife would be too busy helping out with the Neighborhood Association, with the Greater Japanese Women's Association, with fire drills, with packing up care packages and writing letters to soldiers on the front, with taking care of children, with standing in line with ration coupons only to find there is nothing to buy at the end of the line. If you go to a comfort station — if there are any in the neighboring areas — you can't possibly fuck Japanese whores. Off limits: officers only. Why would you pay for whores when there's all these *Chinks* around? Just rape them. These *Chinks* could be enemies anyways — you can't tell who they are, they could be Communists, Nationalists, they're all the same anyway. Just fuck them, they don't care. They are used to it anyways. They don't feel anything — they might actually enjoy it. And once you're done, just kill them afterwards — tell the commanding officers that these women were enemies. That will increase the number of the enemy killed to submit to Central Headquarters. They will nod approvingly and will congratulate the regiment for a job well done.

February 1942: Rationing of Clothing

We are given 100 points if we live in cities, and 80 points in countryside. A piece of underwear costs 5 points; an undershirt 10 points. A dress 20 points and a coat 63 points. What we take for granted, this small pleasure of adorning ourselves with what we want to wear, how we want to be seen, is taken away, but we tell ourselves, *this is all for the war, we are doing this to win the war;* we let them take our pleasures away, one by one.

February 1942: Shirō Teaching at the Young Patriot's School

For three hours a week, he teaches young men civil science and shooting. He shouts at them as he did at his men in China, he tests them on their agility, their focus, their ability to follow his orders, and they do, earnestly, with all their hearts, because here is a man who used to be an Emperor's Guard, here is a man who comes from one of the best families in the neighborhood, here is a man who rose to the rank of sergeant. They listen to his every word, they listen to his stories in the battlefields, they listen to the words of a demi-god, the chosen one, and they dream of becoming like him. In their eyes, he is brilliant, a man of the world, he has gotten out of this land, of this country, and has seen the world, and here is proof, they think, that a man does come out of the war alive, intact, proud. As Shirō scans these eager faces, he feels a twinge of sadness: these boys, so eager to please, so earnest, so hungry for something beyond the familiar perimeter of their lives on the land, too, will go to war when they turn twenty years old.

A Confession of a Torturer

The day's work begins: the military police bring the prisoners in, three *Chankoro, Chinks,* their arms bound behind their backs. They all look like spies, their eyes shifting nervously from one to the other, they make beastly sounds, like sheep or goats, and because they are animals, they must be treated as such. Or they are all spies; why would the Special Police bring them here in the first place? One is left and others are taken away; they will get their turns, eventually, and the waiting will make our jobs so much easier. They will wait in their cells, and they will hear this one scream and beg, and the imagination alone will break them before we touch them. But now, we must tend to this one. He is already pleading, begging, and without understanding Chinese, we know what they are saying — they all say the same thing: that they have a *Good Citizen Pass,* that they pay taxes, that they have old parents waiting for them, they have a wife and children who would starve without them, they are just farmers, they were just walking down the road when soldiers came and dragged them here... all the stories they tell, they are so alike sometimes we think that it is a rehearsed line, taught by the Nationalists, the Communists, whoever we are fighting against. Now, he is making a bleating sound. We drag him over and tell him to lie on the board; he looks at me and pleads, *No, no, please....* We have heard this before. If he cannot understand the language of men, then we must treat him like an animal. After few punches and a streak of blood runs down his nose, he is submissive. He lies on the board and we fasten him down. He keeps looking, staring, pleading. We fasten his head to the board with the leather belt. We take a cloth and gently place it over his face, as if he is a dead man and we are preparing his body for the journey below. He might already be dead, we think, and we push that thought away. Then, we begin to pour water down, so that the wet cloth will stick to his face and he will only be able to take small breaths, so that water will enlarge his stomach and clog his lungs, so that he will confess that he is a spy, that yes, he works for the Nationalists, he works for the Communists, it does not matter which one, and we can move on to another body, then the other, so that we can do our day's work.

1942: Voices of the Rapists — Mythology

Those *Chinks,* they would do anything to survive, they would open their own legs and beckon us to rape them; if it were Japanese women they'd rather die than be violated. That shows what low morals they have.... It was war, when you are cornered, when you don't know whether you are going to live or not, things don't matter. You do it, and in ten minutes, you don't care what you've done, you've moved on to other things.... If they didn't want it, they would have fought harder...after marching for twenty, thirty kilometers without rest, carrying fifty kilograms of things, you don't have anything to look forward to except for drinking or sex...we were all doing it...they'd cut off their hair, they'd dress up as men, they'd smear shit all over their bodies, but that didn't stop us, men need women, you wouldn't understand it unless you were men yourselves...those *Chinks,* they'd walk around wanting us, they all wanted Japanese cock, and we gave it to them....

March 1942: The Disappearing Foreigners

They are disappearing, one by one, loaded on the exchange ships, even those who have been here for thirty years, forty years, those who escaped from the capital of Russia with everything they could carry in their small bags, trekking through a cold land that rejects men, finally reaching a small island nation with nothing in their hands but past glory and present misery; those who fell in love with the country and a woman and stayed here, rooted, seeing themselves as Japanese though they knew that they could never be one in a land that focuses on blood and nothing else. And the words they brought to Japan, *baseball, jazz, pane, dance...* words of the enemy tongues are changed to Japanese words. The good Sisters and Brothers are divided, Germans and Italians can stay, and those from the enemy states are herded into trucks and taken away. God the Father does not care, the Church was not defined by nation-states, but it is now. Some have evacuated to Nagano, where their summer houses are; some have been forced out of their jobs at universities for their languages, the very thing that earned their wages are what bans them from their jobs. When they walk down the streets, it does not matter whether they are Germans, Russians, Americans, or British; people stare at them. Mob them. Refuse money, refuse help. Their neighbors see them as potential enemies; their children with half-Japanese blood are taunted at schools and they come home crying, asking the question, *Who are we? Are we not Japanese?* They evacuate with their embassies; though they have made a choice to live in Japan, some of them as Japanese, it is their blood that defines them, that holds them, and in the end, protects them.

April 1942: The Rifle

He stares into the scope, takes aim, then pulls the trigger, bracing himself for the impact of the gun as he has been taught, as he has taught his men, *Brace yourself so that the barrel does not jerk up and fire off target.* A fraction of a second later, feathers explode midair, and a goose, with its wings open, falls onto the earth. Shirō watches Seijurō run toward it, squealing, *Father, you got it! You got it!,* in the same way a man had laughed as another had sniped an innocent civilian from atop the wall, an old Chinese man falling to the ground, spilling his wares. But he was not dead. He was shot on his thigh, he started screaming, screaming, *I've been shot, help me, help me!* His old wife was screaming, but everyone on the street gave the old couple a glance, but hurriedly walked away, leaving the two men atop the gate laughing, *You missed!* Seijurō comes up to him, holding the goose by its leg, its head hanging limply, *A clean shot,* Shirō thinks absentmindedly; his son looks up with a grin, and he remembers that he is not in China. He is home. He tells his son as quietly as he can, *Why don't we take it home, we'll have goose hot pot tonight?*

April 10, 1942: The March in Bataan Starts

The Imperial soldiers are drunk in their victory, how the white men give up so easily, they tell themselves, how they put down their arms. *A bunch of hired hands,* they snicker, swelling up in their god-sanctioned war, their god-given courage. *They don't know what war is, look how those Yankees ride around in their jeeps whereas we walk, walk, march, carrying everything we need on our backs.* They think to themselves, *Why should we give them food and water when there's not enough to go around?* There's hatred, but it's not a personal hatred. It never is. Just like in China, just like in basic training, the Imperial soldiers take out their frustrations, their anger, on weaker ones, and here they are: Americans and Filipinos, especially Filipinos. The Asiatic brothers who turn their backs against the holy war, refusing to be liberated from the Western imperialists and remaining the faithful dogs of the West. It is a march, a long march of 98 km. *We can make that march in two days, we already have.* There are 75,000 men — both Americans and Filipinos — at the start, but Headquarters knows that one-third of them won't make it. POWs fall from heat exhaustion, from lack of food and water under the brutal sun; the Imperial soldiers rush by in trucks with their swords drawn, cutting throats as they go by, or beating them, just for fun, just like little boys taunting cats and insects, or running them down. A game. This is a game they learned in

China, and how easy it was then. Somewhere around 600 to 650 American GIs die. What the post-war myth does not reveal is that 5,000 to 10,000 Filipinos died, though no one kept count, not the Imperial soldiers, not the Americans who promised Filipinos citizenship if they joined the US military, a promise that was not kept.

April 18, 1942: The First Attack on the Mainland

A clear day. And it comes. We look up to find sixteen planes in the air, and at first we think them to be ours, our own, gallantly cutting through the sky with the rising-sun symbols on their wings, until they start dropping bombs, one by one, quickly, and our hearts stop, we run as quickly as we can, forgetting all about the fire drills, about what to do in the case of emergency, we run to save ourselves. Where did they come from? Where are our gallant men to protect us? We ask these questions again and again as we make ourselves as small as possible, hiding out, and once the bombing is over, the strafing starts, their machine guns chasing after us, bullets flying all around us, and we hold our breaths as if it will make us smaller, make us invisible. *Isn't this against international law? We are noncombatants!* And before we know it, it is over, the planes flying southward, crossing Japan. The radio tells us that fifty people have been killed in the air raid, and one boy was strafed to death, and the government gives him the Buddhist name, *Unfortunately Gunned Down Good Man;* the voice on the radio also tells us that nine planes have been shot down, more than half, and one of them was shot down by a noncombatant plane, and how we sigh in relief, telling ourselves that we are invincible, we will not bow down to the soulless enemy. But many of us wonder, *We didn't see any planes shot down, we know what we saw, there were no friendly planes in the dog fight,* and some of us joke, loudly, *Imperial Army shot down ku-ki,* "nine planes" or "air," a pun.

Mid-April 1942: Private Kōkichi Shimura of the Famed 2nd Regiment Leaving Home

It is late. Since it is what men do, they have been drinking all day long, and these are two brothers saying farewell. It is perhaps the last time they will see each other. Shirō will tell him all he knows about the army life, he will tell all his secrets so that Kōkichi can survive, so that he will keep living. But perhaps Kōkichi already learned them during the three months of basic training; Kōkichi was adaptable, but his surface flexibility hid the strength, the stubbornness, and how Shirō wishes that Susumu has learned to hide his anger, his irritation, his stubbornness. Three brothers with different personalities; three brothers with different fates. Kōkichi has been telling one funny story after another about his time in basic training, and Shirō has been laughing at the way his stories twisted and turned, exaggerated in some places, self-deprecating in others. But it is late, and night forces them to hush down, to keep their laughter low, to burrow deep into their hearts, and the pause between stories becomes longer and longer until they sit quietly sipping *sake*. The light is given by one lamp, creating more shadows on the walls, on their faces, and each gesture is a play cast on the wall, living out an unknown story. Maybe it is the shadow; maybe it is the stories that are not told, but Kōkichi looks straight into Shirō's eyes, his face morphing into an older man's, and Shirō thinks, *This is how he is going to look when he is fifty and when we can sit like this again and laugh about this time together.* Kōkichi whispers, *Take care of my parents. I know that I'm the third son and I have no responsibility — not like you do — but please look after them. Big Brother, don't let them take you, too.* Shirō starts, *This is the time of total mobilization...* and he stops. He sheds the man he is supposed to be, the veteran; instead, he is a brother, the first son, the father of two children, an ordinary man. *I don't think they'll take me — they've already taken you and Susumu, and two out of three is as far as they're going; besides, why draft me when there's so many young men around....* Kōkichi shakes his head, *I don't think all these victories are going to continue....* And the soldier part of Shirō wants to punch him, to tell him that he is unpatriotic, that he is speaking blasphemy, but the man part of him speaks, *I know, the newspapers and radios don't always tell the truth … like I told you, do what they tell you to do, even if you don't agree with them, just play their game, let them like you, go to officer's school, climb as high as you can up the ladder; I'll look after the land, the family, so don't worry about us; look after yourself, and don't waste your life needlessly.* And once the truth comes out, his mouth moves easily. He finally says it: *Don't waste your life needlessly.*

April 26, 1942: Kōkichi Arriving to Busan

It is like Japan, but it is not. A strange miscommunication of the senses: the air smells of fish, of garlic, of dirt and steel; the ear catches familiar sounds, but it is as if he is dreaming of a land where the language he understands is no longer understandable — familiar, but not quite; women look the same, so do men, so do children, but there is a formal distance between him and the people, even when he smiles at them. They do not see *him*, they see the uniform and what it stands for — the Great Empire of Japan, and the burden they have been experiencing for the past thirty years. *It is a part of Japan*, Kōkichi remembers teaching his students, but he can see that there are divisions: Japanese are treated well, and Koreans are second-class citizens. Koreans who speak Japanese are somewhere in the middle. Every sign is in Japanese but men and women speak at the level of a six-year-old. A forced language. Forced ruling. The tension. Kōkichi can feel the tension, the hatred, as he walks down the street of Busan.

Afternoon, June 10, 1942: The Newsflash from the Navy Division, Central Headquarter

Newsflash. Newsflash. *A battle in Midway, our casualties: one aircraft carrier and one cruiser destroyed, another aircraft carrier damaged greatly, and thirty-five planes did not come back.* Their first lie. In truth: four aircraft carriers, one heavy cruiser, one cruiser, and two submarines, completely destroyed; 3,200 casualties, which included some of the ace navy pilots. The first lie. More to follow. But we do not know this; we drink each and every word, and believe that the god of victory is on our side. A mythology. Every character in a mythology is heroic, and we are drunk on this grand emotion.

The Myth of Yamato Spirit, or the Japanese Spirit

Selfless. Sacrificing. Earnestness. Pure heart. Innocent. That's who we are, that's what the voice on the radio calls us, calls our soldiers fighting in the sacred war. One hundred hearts beating as one. Patient. Good-hearted. Resilient. The radio praises us, tells us that we are doing a good job because we are following the Japanese spirit, not the decadent Western spirit. Yes, we tell ourselves, we are our better selves now, we are heroic in our suffering, we are kinder to others because we know they are going through the same suffering as we are: lining up for food, giving up our savings, offering our men, we are all the same, for the first time in such a long time, and we can be our better selves while the dark part of us, the parts that make us human, unheroic, ordinary, are pushed aside.

July 1942: Masa's Pregnancy

He has been back nearly a year now. The war is far away, at least *his* war: the nightmares, the sudden panic, the sudden distancing. It is the battle time on the farm, with the rice crop that needs so much tending. Anything can happen: a hurricane that floods the river, that breaks the dike and washes away the crop. Unruly insects eating away at the green ears. So he keeps an eye on the weather, hoping that typhoons do not come. Seijurō is now six; he follows Shirō around, imitating him, and he sees so much of himself, so much of his wife, in this child. And his daughter is one year old, a sturdy girl. Already stubborn, but looks so much like himself. *A girl who looks like her father will have a charmed life, the saying goes, and he knows this girl will have a charmed life.* When he married, when he was writing those letters, one sheet at a time, through basic training and during pauses in battles in China, during times when he thought himself unmoored, disengaged, he was writing to his wife, his heart, *one heart in two bodies,* he was living for his wife, he kept fighting for his wife. But now that he is back in his life — *his own* — he never knew, until now, until he came home, that there was something more to his life, that his life had a lot more weight, more love, more aches, than he thought possible. His wife. His son. His daughter. These are the people that define him, that give meaning to his life, that ground him in his body, in his life. He feels it. He *knows* it, as he knows that his heart is in his chest, that this is a family. And now, Masa tells him that she is pregnant again, and he is happy. All is well. Life is good. He can ask for nothing more than this life he has.

August 20, 1942: Kōkichi in the Officer School

Shirō sighs relief as he reads the letter: Kōkichi has been nominated and accepted into the Officer's Training. In a war, in any war, being an officer has a more likely chance of survival than being a private. Out of the officers, a noncommissioned officer — a sergeant — has the highest chance of dying; out of commissioned officers, a lieutenant. The more stars you have on the uniform, longer you will survive. How many times did he have to lead his men in China, leading *them,* being the one in front, while his commander stayed back, yelling conflicting

IMAGINARY DEATH

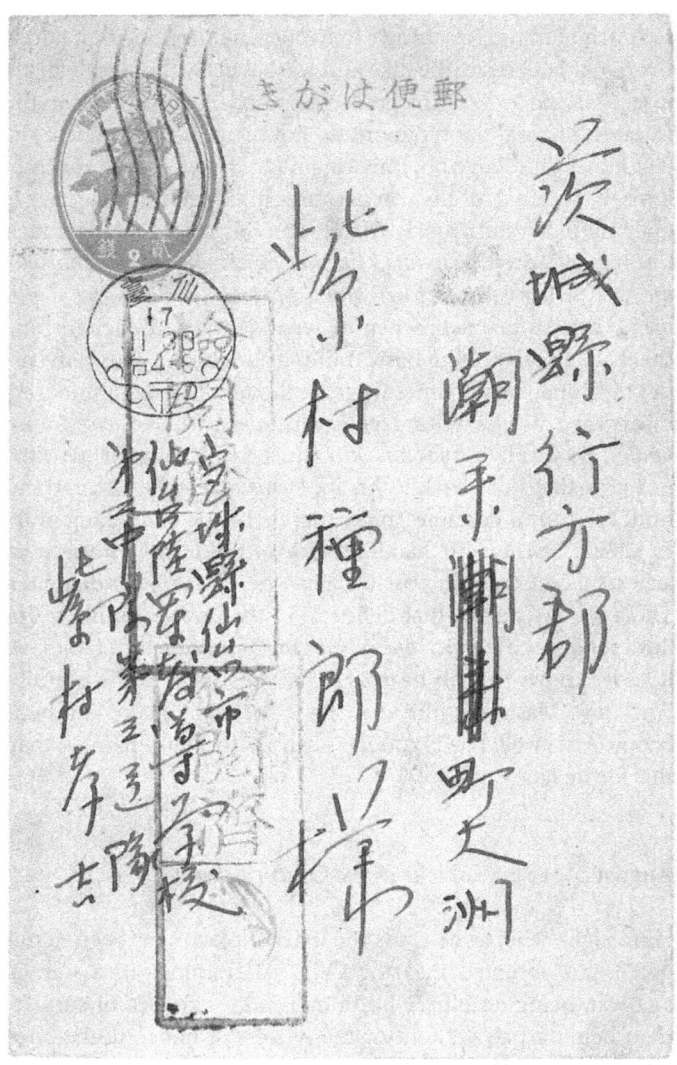

Fig. 4.3. A postcard from Kōkichi to Shirō (front).

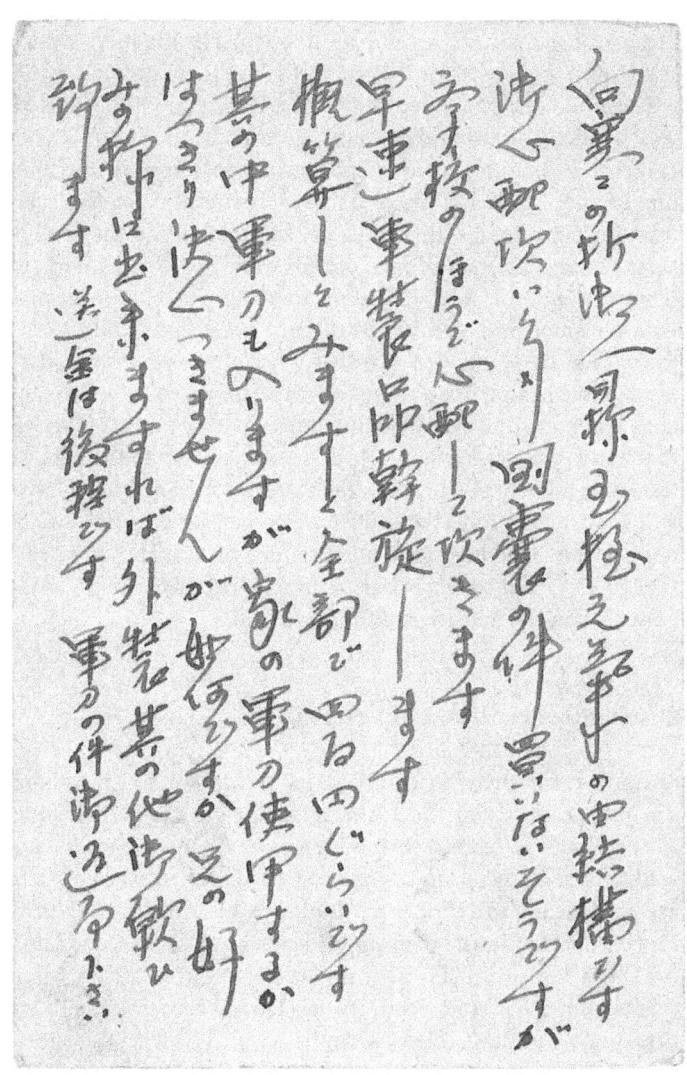

Fig. 4.4. A postcard from Kōkichi to Shirō (back).

orders; how many times did he have to slap his men, punch them, pull them by their collars, just to get them moving, while his commander pretended not to see. How many times did he watch officers sit around drinking, getting drunk, sleeping with Japanese prostitutes, when his men had to stand in line after line for a quick fuck, just so to release themselves for five minutes, for ten minutes, their idea of sex molded into the level of beasts, and how many times did he want to tell his men, *Sex is more than that; it should mean something, it should be done with a woman you love,* but he did not because he understood, more than he wanted to say, that when you spend your days treated like beasts, like a thing, worth nothing, you want to take it out on someone else, someone lower than you. Shirō knows that hands are bloodied. If Kōkichi gets promoted to the officer's track, his life will be bound to the State as long as he lives — but whose isn't, anymore. *If our lives are bound, we might as well be of a value, worth more than others,* Shirō thinks. He sighs relief that Kōkichi will have a better treatment, better pay and food, a better life in the narrative of the State.

October 10, 1942: Two Brothers in China

Susumu is mesmerized by Shanghai, a maze of a city, divided up into areas. From the Chinese section to the British one, to the Japanese section to the French one, Susumu walks and walks, though he is not supposed to. Chinese women here are so delicate, so refined, he thinks, as he walks through the French quarter, music drifting — maybe a violin, maybe played by a beautiful woman — as if this is not the time of war, as if this is all that he read about in magazines. He remembers the first time he went to Tokyo, when he went to Ginza, with short-haired women walking so carefree down the street, with all the imagined and unimagined goods on sale, and he felt like such a country bumpkin. But here, in Shanghai, where people can ease into one language to another smoothly, French sentences followed by English followed by Chinese, one tongue speaking four, five languages fluidly, he can see how provincial he is, how provincial Japan is, although he does not say it. He still does not like the army; he is stubborn in his refusal to play the game, of becoming like *them,* but one thing he is grateful for, that he can see all this, and more. He does not know that Kōkichi is up in north, by the brutal Korean border, huddling in the freezing landscape, but gleeful that he has made it to the officer school, that he was going home in a month. Two brothers in China. Two experiences of the war, and they have not really seen combat, not the way their older brother has.

October 1942: The Unangans of Attu Island of Hokkaido

This is not the snow they know. This is not the village they know. This is not the island they know. When white men came from the west across the frozen ocean, they made the Unangans into slaves; when white men came from the south, they brought with them a white god. The Unangans endured because they knew that time is elastic, and there is more to the weather, to the sky, than what the white couple told them as they stared at the sky with their strange instruments. And then the yellow soldiers came, soldiers who came from the southwest from a small island who almost looked like them but not quite. And who spoke in tongues so unfamiliar, who did not know a thing about snow and cold and winter. With their firing sticks, they told the Unangans to go inside the steel whale, and until then, they had not known that the ocean was so big. It took them so many sunrises and sunsets. All forty-two of them. The entire village. They took the villagers to an island called *Hokkaido*. They handed them shovels and told them to dig. *We don't know how to dig, we know how to hunt seals and how to tell big snow from small snow, how to skin animals, but we do not know how to dig,* the village leader protested, but the yellow men do not hear. They shake their heads and gesture to dig, dig in the cave. So they dig; half of them get sick; they are ordered to dig, no matter what. Nearly one-third of them die during the first winter.

What is war? they ask amongst themselves; whose war is this but a war between yellow and white men. This is not the war the Unangans have started, but they are captured as prisoners of war. They will remain in Japan until the end of the war, and when they are freed, the ship will take them southward, away from Attu island, away from their village, and the ship will land on the southern shore of Alaska, away from their home. Even when the war ends, they will not be able to go home.

1942: Gods Do Not Approve of Laughter, but There Is Laughter

We must be solemn. We must be earnest. This is life and death, this is not a joke. War is not a laughing matter, you don't see our Emperor–god laughing. You do not laugh when you think of your men, *our* men, sleeping in foxholes and trenches, crawling through mud and walking in snow, so why should you laugh during your everyday life? Laughter is banned from public spaces: theaters, comedy halls, cafes, restaurants, on the streets. We walk about with our faces solemn, carrying the gravity of the war, carrying the fate of our nation, the seriousness of this era, which is deciding the next hundred years of this nation. Plays are censored — which lines are permitted, which actors can be on stages, which plays can be put on. Chekhov, no. Shakespeare, no. Kabuki, yes. Comedies are abolished: What are comedians supposed to do when their words are taken away from them and there is no audience to laugh at their jokes? We do not laugh on the street — we never know who might report us to *Tokkō,* the secret police, who come at all hours of the day to take us away for being unpatriotic. The only laughter allowed is that of the soldiers far away on a different continent — the nervous laugh of soldiers as they kill the enemy, with bayonets, with rifles, with swords; they laugh not because they enjoy their work, but because they are revolted, because they are nervous, because they are scared. And the Chinese children and men and women, those who somehow will survive the war, will always remember the laugh, the devil's laugh. *I can't forget the laughter of the Japanese-devils as they beheaded my father, as they raped me, as they....* And laughter will ring out in their dreams and in their memory. Our men laugh because they cannot protest, because they are revolted, because their bodies betray them, and the only form of protest is the laughter that comes out inappropriately.

November 27, 1942: We Will Not Want Until We Have Won

When a soldier at the front received his salary — every ten days, if all was quiet on the front — he only received one third of what was his due; the rest was forcibly withheld in a postal savings account. That postal savings book and his insignia stamp are two of the most important things he carries with him through the battles. Women back home bought war bonds and saved as much money as they could, because hamlets and villages forced their people to put money into a collective account for the war effort with dark threats that if people refused, they could have ration coupons withheld and a silent censure from the farmland, where no one would come to help them till the land, to help with the harvest. In the total war, men were forced to finance the very war they were fighting in, with or without their permission, and women were forced to finance the war that their men are fighting in. And a slogan was introduced: We Will Not Want Until We Have Won. This was the slogan of the year; everyone was equal in hardship in a way the nation wasn't before the war.

1942: Dogs, Cats, Rabbits

No animal is safe. Not the old horses that must be drafted. Not the carrier pigeons that shoot through the sky, carrying messages. Dogs have been disappearing, and now cats. For meat. For leather so needed by soldiers, though no civilian can find leather anywhere. And rabbits. Of course, rabbits. For meat. For fur. No animal is spared. Not for the war effort. Never for the war effort.

December 1942: A Reunion at Kashima Shrine

A reunion between two soldiers, a lieutenant and his former noncommissioned officer. A lieutenant from the China battlefield. One of them, still an officer, and the other one, a civilian now. Shirō is relaxed. Masa is about to give birth to his third child, and he is happy. Seijurō is already in first grade, and does not shy away from him, as he had done when Shirō first came home a year and a half ago, and his daughter, Yoshiko, nearly eighteen months old, walks already. Yes, and talks. It is a meeting, a very brief one, but the lieutenant and Shirō talk of the good old days, of shooting geese over the frozen pond where they had built the barracks, of China, of having survived, both of them, though one of the lieutenant's legs is an inch shorter than the other from the shrapnel that pierced his left thigh. Shirō talks of his family, of his pregnant wife, of how content he is. His face is not at all tense like it had been back in China. And it is time to go, they tell each other, *Take care of yourself.* Back home, there is a phone call to the village office, *We need more men.* He is about to be drafted again, but he does not know this yet. The news will arrive as a surprise, as it always does; at this time, he thinks himself safe, though he has a sense of foreboding.

December 14, 1942: Susumu in Rabaul

When they left Shanghai on November 21, they were not told where they were going. They were not told whether they were going north or south, or back home, but as the convoy kept moving through the ocean in its usual zigzag formation, as they sweated in the cargo hold of the ship, the temperature rising one degree at a time as hours turned into days, and they sweated, sweated, and were cursed off board when they went to get a fresh breath of air, they knew they were going to the southern islands. Which one, they didn't know. Rumors start, because they are bored, because soldiers are good at it, because they know they can't trust the official words of the state, of their officers' words: *Guadalcanal, New Guinea, Solomon Islands.* Where are these islands? No one knows. *Burma. Thailand. The Philippines.* Audacious names from geography books come out, but they know: their winter coats and uniforms have been taken away, and they have been given summer combat fatigues, that strange color that is not quite olive, not quite khaki. They go up and down the nets, they run up and down the ship, they are told to unwind their loincloths if they are thrown overboard so that the cloth will trail after them and the sharks would not attack. It was relatively quiet, though they could sense the tension of the Navy men. Someone whispers, *I heard that the Battle of Midway was a disaster.* Someone tries to hush them, but the other continues, *I heard from a buddy of mine at the Central Headquarters that they lied about the casualties. Heard that this sea is infested with Americans and Australians.* Susumu listens to it as he hears the names of southern islands. Not quite believing, not quite disbelieving either. And two days ago, he landed here on this island of thick green, the sea so clear he couldn't tell where the sky ended and the sea started, where the volcanic mountain still smothered from heat when it exploded back in June. They landed without incident. Their regiments were divided up into companies, then into platoons, and told to go into the jungle to learn how to make shelters, and they did. They cut down the stubborn trees, they drag the logs—because there are no trucks or horses here—and assemble them into crude shelters. Then they go around shaking coconuts from the trees. They dive into the clearest sea, and Susumu swims like a dolphin, amazed that the same sea up north near his home is harsh, unforgiving, but here, it is calm, clear, and beautiful. For now, this island is unreal, almost like they are on a camping trip, leading a life of Robinson Crusoe, just like they all have read when they were boys and glorified adventures, so they sing, they get drunk, they dance, just like boys on adventure, until late into the night, where the parrot screeches, unidentified animals howl and sing, insects create music never heard anywhere, and the darkness, not yet malignant, not yet exacting.

December 20, 1942: His Last Child Is Born

A man is not supposed to be present when a woman is giving birth; it is considered bad luck, inauspicious, and a woman giving birth is dirty, impure, must stay away from the cooking fire, from the shrines besides. But Masa didn't quite make it back to the birthing room — a shed — where no one could hear her; he had to hold her as she screamed, he had to lead her to the room as his mother took over, as she shut the shed door, shutting him out. He went back, holding Seijurō's small hand in his. He did not care whether it was a boy or a girl, he only cared that the child was healthy, that this child stays alive, to lead a charmed life. For Masa to stay alive. How many women have lost their lives for giving birth and not recovering after the birthing. The quiet of waiting. Minutes. Hours. The first cry. Then a wail. Shirō cannot help but to get up, to walk toward the shed, and Seijurō trails after him in curiosity; the door remains shut, and Shirō presses his hand on the wooden door, closes his eyes. His mother comes out of the shed, holding a bundle, holding a bundle the size of a cat, wriggling, still crying. And he peers in: a baby. A girl, his mother says. A baby, her face still red from her journey down the canal, her hands in fists and her eyes shut. He has seen so many men die in front of him, he has heard their last exhale, has seen how their souls depart from the bodies, but this. A birth. He reaches over, almost afraid to touch this thing, this small bundle of a child, with all hopes and desires and dreams, and he cannot believe how this child will turn into a person eventually. And this child is his and Masa's, conceived during the winter month of February. He reaches over, and his mother hands him the child. A jolt of surprise, sadness, beauty, happiness, rushes through him. A miracle, he thinks, though he knows that it is something everyone says, but what other word can describe this beautiful human being, who trusts all of her weight in his hands? Seijurō peers in and smiles. Shirō has not cried in a long time. Now, for the first time in his life, he cries from joy.

December 27, 1942: A Strange Look

He pedals through the familiar route, the very one he used to take when he was still in school. That was before the war. That was before the marriage. How he would bicycle as fast as he could, a forty-minute journey through the rice fields, across the river, until he could almost smell the ocean, but not quite. But today, he is on a different errand: to file a birth certificate for Matsuko, *a pine child, a*n auspicious name for his third child. The village office will close for the New Year, but he could not file the birth certificate, not until seven days, making sure that his child survives its first seven days on earth. But then, even then, there is no guarantee that she will survive the first thirty days, then a year, then three years, five years, and, finally, the seventh year, when she no longer belong to the gods and the otherworld, but to this earth, to her body. But Matsuko is a healthy child; she takes milk vigorously, and Masa is doing well as well. The office sits in a small building, and as he enters, the officer addresses him as Sergeant Shimura. A renowned veteran, a salutation fit for the former Emperor's Guard. But the air is tense; there is something strange about the officer, who he has known ever since he was a child, the one who was one of the first ones to bring the good news when he was chosen into the Guard Regiment. He does not meet Shirō's eyes, no matter what. *Is everything alright?* Shirō has to ask, but the officer just minces his words, *Oh, it's been busy here....* Shirō tells that he has a new child, and that they've named her Matsuko. The officer's face darkens for a fraction of a second, and the reply of congratulation is not as hearty, but weak. Shirō does not know what to think of this; he tries to shake it off, and walks away. He did not see that there are files and files of village men, all between the age of twenty to forty, opened; some already stacked in a corner to signify that they are already drafted and serving. The ones open are the ones that will be drafted, very quickly, just after the New Year celebration. A clerk opens the file: Susumu Shimura drafted out of the tax office in 1941; Kōkichi, conscripted, and now in the officer's school in the north in 1941. The Central Headquarters needs a sergeant, a seasoned drill sergeant to train the new recruits, and they must be able to do it now. Shirō Shimura. It doesn't matter that he has already served two years in northern China, crawling in the mud and being in battles and learning how to kill or at least to disarm the enemy. And it doesn't matter to anyone that he was chosen as an Emperor's Guard, the elitist of the army unit, when he was twenty years old, with his wife pregnant, with a child on the way. It does not matter that he is the eldest son with a family of his own. In 1943, he ceases to have a military history in the eyes of the government. He is only a quota to be filled.

January 5, 1943: The Draft Notice, or the Dreaded Red Notice, One of 960,000 Men Drafted This Year

The notice came right as the year turned new, with a message telling Shirō to appear at a certain station at a certain date — January 5. He was told to dress in a civilian outfit so that the enemy would not make note, *after all there are so many spies here, we can't let the enemy find out the mobilization of our men.* Told to come alone, without the fanfare of the nation. He knew it would come, he waited for that bicycle from the village hall, but he did not know that it would be so soon. They had already taken Susumu in March of 1941, just as soon as he received a degree from the university; a year ago, exactly a year ago, Kōkichi was taken. He was the only one left, and his parents thought that they would be spared at least one son, especially the one with three children. Yes, he was now a father of three. Who would look after the farm? Who would take care of the rice paddies and his aging parents, who would look after his own family — his wife and three children? But he must go. He wakes up early, before the sun is up. Masa helps him put on his uniform, packs up his lunch with the meager supply of food. They do not need any words; there is no word needed to explain all that needs to be said. He loves her. He will come out of this alive, as he has done in the previous tour of duty. He also knows, but doesn't say, or even let show in his face, that Japan is desperate to call men like him to battle, that with the way things are going, with so many men being pulled into the war, they needed a retired sergeant like him, to whip the new recruits into shape quickly, that they will be running out of young men — they already have. He tells Masa not to wake the children, but he goes over to where they lay asleep, touches each child's cheek, one by one, from Seijurō to Yoshiko to Matsuko; how he wishes that he will be able to do this when they are older and the war is over and he is back home. Masa is watching his every movement. He nods. He pulls his rucksack over his shoulder. She follows him out of the dark house, quietly, as they used to do when they made love amidst the sleeping bodies of his family. They hold hands as they walk slowly, trying to stretch the goodbye, trying to wait until the very end, until the edge of the village, where she would stay and he would keep going. His last words to her: *Take care of the family, look after the children's well-being, look after your own health. I will write often.*

Fig. 5.1. Gorō, Masa, Seijurō, and an unknown boy.

V

The War without an End

January 5, 1943: On the Train to the Unknown Destination

Men, all wearing war-time colors, the civilian uniform called *defense clothes,* stand on the platform, waiting for the train to pull in. No one talks, not the way they did when they were going to China so many years ago, but this time, there is no fanfare. Men were told to leave as quietly as possible, never to tell anyone when the train was leaving. *You don't know where the enemy spy is,* the warning went. And looking around, he sees that there's almost no young men, not like the men he trained back in China, their eyes bright, their faces full of anticipation of the possible heroic deeds, their bodies straight in earnest hope for the bright future; these men Shirō sees are older, tired, with the knowing look on their faces: *we survived the first time, what is the likely chance I will come out of this alive?* That's what Shirō thinks, too. He so wants to see his children grow up; he so wants to hold Masa in his arms, to grow old with her. But the Emperor has called him in this time of need, and he consoles himself with a year and a half of returning into a man, a husband, eighteen months of dreaming about the future, of living like a man. The train pulls in; the shutters are all closed, and they sit quietly, strangers to each other, although they know what is to come.

What is to come: the army chose the veterans, men they don't need to train, men who are combat-experienced; will drop them into the middle of battles, and they will shed their humanity, the very thing they tried to keep in their previous service, the very thing they tried to find, to hold on to during their time back home. Shirō sees the familiar landscape blurring through the tiny cracks between the shutters: deep yellow, wintering earth. He looks away. He lights a cigarette, then offers to a man sitting across from him and starts the familiar conversation of, *which neighborhood are you from?* The train keeps chugging through the quiet landscape.

Mid-January, 1943: Manchuria

The boat ride was the same, packed, sweaty, but unlike the previous time, when men knew each other, forming a brotherhood already in the basic training, this time, men, hardened and cynical, sat in their 80 cm of private space, and complained about this or that. From Moji Harbor, only a day's trip to Busan. Sharing battle stories. Sometimes boasting, sometimes cruel, sometimes exaggerated, but most of the time just a story told without responses from other men. There are no bright eyes here; there are no flushed cheeks, no blind faith in the Holy War. That is for the youth; what the government says can no longer be trusted. In China, things are getting worse and worse; it was supposed to be an easy war, but the Chinese, the Nationalists, the Communists, those *chankoros* would not give up. They knew they were going to China; they've been given the thickest coats, sizes varying, furs around the collars frayed from previous owners. They spent the short boat ride cleaning their pistols and rifles and rusty bayonets; they smoked packs after packs of cigarettes and drank as much as they could. Probably the Soviet border. Faces etched with deep worries, with a lifetime of aches that can only be seen in shadows. They were all thinking the same: we have survived the previous time, but will we survive it this time? And as they disembarked at Busan, they were put on the open cattle train where they had to piss from the slits, where they huddled together in groups to keep warm, snow falling on them mercilessly, through the snowscape to Hongshui in Hebei, with their rifles ready to fire, with pistols within easy reach. To the west. So not to the border, but to the midst of the fierce battles, where peasants, rebels, women, old men, all were *chinks*, spies, to be shot on sight. And as soon as the train stopped, they were redistributed to their new units to replace the dead: Shirō to the 237th Regimental Unit, 5th Company, with a dozen hardened men under him. Most of them from Ibaraki Prefecture, speaking in the familiar dialect of home, just like any regimental unit. A squad leader. An NCO leading men from his home. An instant brotherhood. And it is so easy to get be a soldier; the moment he steps into the battlefield, where men are marching, marching, always alert. He remembers, his body remembers, how to be alert, how to keep his ears open, how to move, how to decipher how close the bullet is, how to distinguish friendly fire from that of enemy's. And when one of his men is shot, he doesn't feel anything, although he should. They do not have a shared history, yet. They are not his brothers, yet. How easy it is to become a soldier, and how he hates it.

1943: The Voices of the Dead

We make our ways to the shamans, the ones who hear the voices of the dead, the ones that will tell us what our men must tell us, their voices from the other world. The shaman will slowly invoke the spirit to come back to this world, to enter her body, to impart the last message, or the messages. She chants the prayers of the heathen god, the one we are supposed to deny because we have a new god, and the Emperor–god has ears everywhere. She writhes; her body resists the intrusion of another soul. She moans, and we rock our bodies in the rhythm of her moans. And suddenly the air cracks, and he is here, our dead one. We ask how he is, and the shaman, in a voice not her own, starts, *I have died and crossed over the river Sanzu, but I am here now....*

IMAGINARY DEATH

Take care of the children, take care of business. You will find... a litany of regrets comes pouring out of her mouth, and we drink up the words because the only proof of his death was the letter from his officers, from the regimental headquarters, from the government, *He died gallantly,* and we are comforted in the way we are supposed to be, but more than anything, we wanted his last moments to be with us, all the messages imparted in the dying hours, when regrets are resolved and messages heard. But when he died, he was so far away; there is no proof. So here, we incline our ears, we listen to the words of his regret, we hungrily take in the messages he must tell us to console ourselves, so that we can carry them with us because he is no longer here.

January 28, 1943: The New Command

The command comes abruptly, in half a day's notice: that 237th Regiment is to be transferred to a new destination, to be announced later. That they must pack up their gear and march half a day back to the train station, and get on the train, south, south across the great Yangtze River, the train not stopping, even through the green zone, and men are told to carry their pistols, to be on alert, to take turns sleeping on cattle trains where they huddle like the cattle that they have been reduced to. Trains stop; shots are heard, and they flatten themselves against the hard biting floorboard, their muzzles aiming toward where the shots are coming from. Silence. Then the train starts moving, and they sigh in relief. Only for the fighting to start all over again an hour later, three hours later. Food is thrown at them from the roof, and they scramble like beggars, almost like those enemy kids they used to keep as pets. The destination: Shanghai. A day's travel, nonstop, from the Yangtze river.

February 1943: Women, Too, Must Fight

The Greater Japanese Women's Committee has started the training in earnest. Procuring vets from each district, women of all ages, children, and men rendered too old or too valuable to be sent to the battlefront all stand with headbands tied tightly around our foreheads to show our seriousness, to show our loyalty to His Majesty the Emperor, to physically make us aware of the focal point. We stand at attention with bamboo spears in our hands, as we have seen the soldiers do. We are ready. *Back, back, front, front.* We move as commanded, taking steps back, back, front, then two steps in the charging position, then *stab*, thrusting the sharpened point at the invisible enemy. We are ready for the war on the home front. If and when the Americans come, we will stab the enemy with our homemade weapons; we will thrust and stab rather than submit, rather than live the shameful life of a captive.

February 4, 1943: Fall of Guadalcanal

The Central Headquarters announce: *From August of last year, the regiments responsible to defend Guadalcanal Island, one amongst the Solomon Islands, have been under heavy enemy fire, but they have been successful in accomplishing their mission by forcing the enemy to retreat; they will be leaving said island at the beginning of February, and will be mobilized to a different strategic location.*[1] The Central Headquarters gave up sending supplies: why waste good convoys, just to feed and rearm 30,000 men? Soldiers are scattered everywhere in Asia and the south Pacific now; they, too, need food, need new men, reinforcements, ammunition. Reality is far worse: out of the 31,404 who landed on the island, only 10,000 made it out alive. The rest were abandoned on the island, without any supplies, any reinforcements, forced to

---

1 Wikipedia, s.v. "ガダルカナル島の戦い," http://ja.wikipedia.org/wiki/ガダルカナル島の戦い.

retreat into the jungle — green desert, someone called it — to go crazy, to slowly starve to death. They were told to defend the island from the enemy, but how could they when there was no ammunition left. And when there was nothing to shoot with, the command came again to fight, to fight until the last man, and they had no choice but to charge — Americans called it the *banzai charge* — to charge to their death with swords, with yells, with their will, only to be machine-gunned down. And the commands kept coming from the Central Headquarters: keep fighting, but be self-sufficient. And men hid in the deepest womb of the jungle; they sat down and could not get up because they had not eaten for such a long time. They first ate dreams of going home, they ate wishes, and when that wasn't enough, they ate everything they could: leeches plucked off from their bodies, strangely glowing mushrooms that made them vomit then die from dehydration, drinking seawater because there was nothing else only to find themselves thirsting for more, and the body, already starved down to the thinnest flesh, with flesh stretched taut under the thin skin, could not stand the smallest infection, the lowest fever, and it succumbed to the slightest violation. A soldier writes on December 27, 1942:

> *This morning, several men went back to heaven, again. Flies flock and buzz around corpses lying around on the ground. It seems like we have reached the very limit of the human body. People who are still barely alive — their faces are the color of the earth, and their hairs are thin and sparse, just like a baby's. I wonder when our black hair turned into downy hair. There's no more energy, nutrition, left in the body except to grow downy hair. I've read in a novel about a human being with sparse hair... but with the energy left, there's nothing left to grow hair. Small-boned people thin down to their bones; fat people bloat. From the gold crowns and fillings falling from our*

teeth, I guess even our teeth are rotting. I never knew that even teeth are alive.[2]

Twenty thousand either died or are missing in action. It is said that nearly 80% starved to death. Men called it *Gatou,* instead of Guadalcanal, they called it the starvation island. But the Central Headquarters look the other way, they tell us that soldiers are now mobilized to somewhere else. They are. They have been mobilized to the otherworld.

February 1943: Shirō in Shanghai

The cargo train that chugged southward from northern China stops here: Shanghai, the Paris of the East, where women and men dance with their bodies entwined to the latest Big Band music, where languages are as fluid as money spent on wine and champagne. He still remembers that letter Susumu wrote from this city, how he praised the cosmopolitanism, how he had thought Tokyo as the center of the culture, but now he knew better. How long ago was that — only five months ago? The letter arrived when Shirō was still home, and he read it to his grandmother, as he used to do when he was young, a job he asked Masa to do when he was in China, and the job he knows Seijurō will take up now that he is old enough, and he has asked his son to do so. Shirō walks through the main boulevards, catching this or that word in English, in French, in Chinese. And when he speaks his childish Chinese he learned on his previous tour of duty, the vendors keep their faces straight; he now knows that Shanghai dialect is smoother, less harsh, more refined than the ones he learned up north. This is the first time he learns of the refugees, *white men* and *women,* with various nationalities and languages, tied by their blood and God: Jews in the bombed-out Hongkew neighborhood, streets with only skeleton of struc-

---

2 Akira Fujiwara, *Uejinishita Eireitachi* [*The War Dead Who Starved to Death*]. Tokyo: Aoki Shoten, 2001.

tures left. These refugees walking across Europe, getting false papers, booking themselves on liners and Trans-Siberian trains to Vladivostok, nearly six thousand miles of journey, to end up here, in the Far East, awaiting the exit visas which will take them to Australia, to America, to, strangely, Japan, somewhere safe, anywhere other than Germany, Austria, Poland, where their people are getting massacred in thousands and millions. They had abandoned everything: their dead, their property, their shame. It did not matter that back home, they were world-class champion chess players, composers, physicians, businessmen, concert violinists. All that was lost along the way to Shanghai. At this time, a Japanese government representative calls forth the Jewish representative, when the German government pressured them to hand over the Jews. *Why does Germany hate you so?* A rabbi answers in Yiddish, "Zugim weil mir senen orientalim — Tell him because we are Orientals."[3] The Japanese government ignores the pressure and puts the Jewish refugees into a ghetto, just like the ones in Krakow, in Warsaw, but without the threat of deportation. Shirō does not know any of this, he just knows that here are Jews whom the Germans hate so much, and he does not understand why. Instead, he takes up a pen and begins his letter to his wife and his children, *Dear Beloved Masa, Dear Seijurō, I am now in — and what a wonderful city it is....*

1943: The Central Headquarters Call for 960,000 Men

This wasn't supposed to happen, not at all. Men who had served before were supposed to be saved, having already served their country and now that they were home, they could have gone back to their interrupted lives. Those who made it back from their first tour of duty. Those men who, at age twenty, were deemed superior enough to be a part of the Emperor's army. Must be healthy. Must be at least 5 feet tall (152 cm). Must be

---

[3] Warren Kozak, *The Rabbi of 84th Street: The Extraordinary Life of Haskel Besser* (New York: HarperCollins, 2004), 177.

of average or above intelligence. Must be of reasonable sight and not be colorblind, though that was changed later on, as the war worsened and they needed anyone. Must stand naked at age twenty, line after line of naked boys, the doctors would fondle their penises and testicles to make sure men were free of sexually transmitted diseases, and next, they would tell the men to bend forward so that they could check the anus by plying it open and checking for any signs of deformity and anomaly. The rest — the short ones, near-sighted ones, and amongst them, men who were too old or crippled, or, in a case of a boy named Kimitake Hiraoka, unfit to serve through lies he told to the doctor: I have tuberculosis. He later on wrote novels and became Yukio Mishima, although he never forgave himself for being not good enough to be a man, and who, in 1971, committed suicide at the Ministry of Defense to prove to the world, or perhaps to himself, that he was man enough, a warrior enough. But that is later on. This is 1943. Men who had passed the conscription physical with level A or B before this time period had already been called to war, to defend the Great Asian nations from the Western aggression, the war to bring about peace to Asia and to the rest of the world. They have done their time and they are now back home, trying to salvage the lives that had been broken into two pieces: before, and after. But in 1943, the draft system, running behind the scene from Central Headquarters, calls forth the numbers of men needed. It is not whom they want, but how many. A town might receive a call in the middle of the night, asking for five men. The town hall clerks open their secret files: all eligible men, fit ones, have been drafted; the remaining men are the ones deemed unfit to serve the first time around because they were first sons, men with families, men who had served already, men who were too short, too cunning, men with questionable ideologies, or men with specialized skills who were needed on the home front; men over forty; boys in their teens; students. But with soldiers dying or made incapable to fight in all four corners of Asia — as well as all oceans in Asia where the Imperial Navy was fighting — they needed more men than they could supply. So the calls came in the middle of the night: We need more men. And the numbers will be fulfilled.

February 30, 1943: Shirō Heading South

They have been on the boat for a week now in the sweltering cargo hold. They have not been told where they are going, and the heat is making everyone short-tempered. Seven cargo ships protected by three destroyers. Just like the prostitutes several years ago, when the sea was still safe and men still hungered for women to fuck. What differentiated them was that prostitutes were labeled *army supplies,* whereas these men are still considered *reinforcements,* but that does not change the fact that whether you are a prostitute or a sergeant, you are held in the cargo like delivered goods. Only once a day, always at night, are these men allowed to emerge out of their hold to get a breath of fresh air. They have stopped at Taiwan for three days, and each man was given a bunch of bananas. Bananas that cost a fortune back home, and never the ones so fresh, so sweet, and they begin to dream of the place they may be going where bananas grow on each and every tree, where coconuts fall from the sky, just like in adventure stories they grew up reading. Now, they are going on an adventure of their own, south, southward, amidst the vast ocean, just like the boy-adventurers in these stories, though they have had their adventures already and know that adventures belong only in books.

March 1943: The Foreshadowing

Masa has not heard from Shirō, who had always been prompt, regular with his letters. Three months. An eternity. She has heard from her youngest brother that, *off the record,* Shirō's regiment was heading to northern China. As Shirō wrote so many years ago, *no letter means that I am serving the country well;* she tries to tell herself that as she goes about the day, taking care of her three children, her in-laws, the grandmother-in-law. At the dinner table, they try not to think about Shirō, Susumu, and Kōkichi, although at every dinner, someone says, *I wonder....* She also knows that her father-in-law wakes up early in the morning to pray to the ancestors for their well-being. She knows that her mother-in-law, at vulnerable moments, rereads the letters her sons had sent. Masa takes down the books of China from the shelf, the one with beautiful photographs accompanying the history of the country — the book that Shirō had brought home in his rucksack when he came back two years ago — and opens the page to northern China, just so that she can see what he must be seeing, just so that she can feel closer to him, even if it's on the pages of a book.

March 2, 1943: Susumu on the Teiyo-maru

Susumu must have been on the *Teiyo-maru* that left Rabaul at 23:30 on February 28. It carried in its womb 1,988 men, and by the time it was sunk on March 3 at 08:00, only 73 men survived. At Rabaul, they were given a dried fish and a float each, *you can survive eating dried fish for many days.* He must have sat in the womb of the ship with others, listening to the whirls of engines, of bombs hitting other ships, the surface of the ocean, of yells and screams coming from the starboard above. He must have known that he would not make it back — men on ships seldom did. He must have made himself as small as possible, knowing that this would not help. He must have been praying like the others when the alarms went off on the night of March 2, and he must have known that he would not be able to swim for his life. That he was no longer an individual human life, but to the enemy, a ship, a creature amongst the brethren of ships. An individual man, with a singular life, could have surrendered with a white flag in his hand. A man could have held a rifle, seeing the enemy eye to eye and saw humanity in each other. But a ship in the ocean. In the enemy's eyes from above, it is a target. Only a target. Bombs from airplanes explode against the sideboard.

IMAGINARY DEATH

履　　　歴　　　書				
氏　　名	柴　村　　晋			生年月日　大正7年10月7日
退職当時の	本　籍　地	茨城県行方郡潮来町大字大洲50番地		
	官　　職	陸軍上等兵		
叙位・叙勲	なし			
年	月	日	任官・進級・昇給	記　　　　　事
昭和16	3	20	二等兵	現役兵として野砲兵第14連隊第7中隊に入営
	7	29		野砲兵第14連隊第7中隊に転属
	8	13		満州派遣のため宇都宮出発
		16		神戸港出発
		20		壺蘆島上陸
		23		錦県着
	9	20	一等兵	錦県出発
		27		錦県出発
	10	1		壺蘆島出発
		10		黄埔上陸
		12		増城着
17	11	21		黄埔出帆
	12	12		ニューブリテン島ラボール上陸
18	3	3	上等兵	ダンピール海峡海上付近において戦死
				以下余白

上記に相違ないことを証明する。

平成19年5月8日

茨城県知事　橋　本　

Fig. 5.2. Susumu's army resume.

One ship breaks off rank as it burns and the enemy airplanes chase after it like flies. And he sits amongst others, praying and praying that he can go home alive.

March 3, 1943: Susumu's Death

If he had survived the initial bombardment that went on for a day, when the Allied planes came in waves, dropping bomb after bomb on the convoy, he and other soldiers cowering in the ballroom of the former ocean liner made into a merchant ship; if he had survived jumping off the ship as he was trained to do in the sweltering Palau; if he had gotten on the life vest in time and trusted that he could remain afloat, relying on the swimming he learned in the harsh and unforgiving waves of the Pacific near his home; if he had swam quickly enough away from the sinking ship so as not to get pulled into the whirlpool that sucked the weaker swimmers in; if he had survived all that, what awaited him was the water glistening with fuel from the sinking ships, where fire traveled so quickly over the surface, furiously, regardless of the water underneath. Where anything above water burned so quickly, turning men's heads and faces into charred masses while their torsos underneath the water remained untouched. Where the sharks dragged men down quickly, leaving a trail of red blood blossoming in the water all around. If he had survived all that, he would have seen the Allied planes buzzing like angry bees — *in waves, like flies* — gunning down men and rafts. Susumu would have watched pillars of water spurting out all around, men all around him suddenly losing the energy to hold on, to fight — energy suddenly leaving them, floating all around him. He would have seen bodies floating by, all in different degrees of incompleteness: arms missing, heads missing, an orphaned leg, men floating with their faces up. They were only a game for gunners: *get those Japs!* If Susumu has survived all that, he would have had to face the vast ocean all around him, where there is no land in sight, anywhere. Alone.

March 3, 1943: The Spirit Travels through Many Oceans

Susumu appears in front of her, dripping wet, as if he had just walked through the ocean. Standing in front of her, he tells her, *I'm so sorry, I have so many things I wanted to do, but I can't anymore. I'm so sorry, I'm so sorry,* and he disappears into the dream, and she does not wake up, no, she is awake. She opens the door, she goes out in the yard, into the farm, looking for him, at every nook and hiding places, but he is nowhere. And she wakes up. A dream, she tells herself. A bad dream, and she goes back to sleep.

March 7, 1943: Shirō Arriving in Palau

One knot at a time, creeping southward. The convoy creeps in zigzag, and with each hour, the cargo hold Shirō is held in gets hotter and hotter. Someone says that it is sunny out, with high waves. It is sunny, he can feel it. The hold smells of urine, sweat, feces, horse manure, of beasts, of dead bodies, but no one has let them out on the deck, *too dangerous*. Instead, they are told to stay down, down. Horses begin to die and they haul the carcasses with a rope and throw them overboard; men are short-tempered, and small things set them off — someone stepping on their legs, someone with diarrhea, someone who got the imagined bigger portion than others — and fights erupt here, there, like landmines. The threat, the unsaid threat, is everywhere, started out with truths that became rumors that became truth: *No ship has gone through this route unscathed. There are sharks everywhere, sharks that can eat you alive. No one survives when the ship sinks.* Shirō can taste fear; it tastes sour on his tongue. The only thing Shirō can do is close his eyes and try to imagine writing a letter to his wife. He must have been at sea for two weeks, the longest he has been on a ship, the monotonous time forcedly forgotten by sleeping, sleeping — but how much can a man sleep — until one day, a cry, *Land!* And the word is repeated from one mouth to another until the entire cargo is buzzing, *land, land, land.* They do not know where they are, only that there is land ahead, and that they are landing.

March 1943: A Soldier's Voice

Whenever I had a chance, the old veteran would say, I would go around and give candies to Chinese children who would flock whenever I went out the gate. They were so cute, reminded me of my little cousins, you know. They would tell me, *thank you, thank you,* and I would play with them. I didn't like to stay in the barrack, so whenever I had a chance to go out, I would take a walk and they would invite me to have tea or dinner, and I would go into their homes and sit with them and talk. I treated them well, you know. They liked me, and I spoke a little Chinese. Sometimes, we had to rough these Chinese up, just enough to scare them, but I never hurt them. The old veteran closes his eyes and sighs in contentment. But what he does not know is that these children do not remember his kind gestures, maybe one or two would say, *Yes, there were one or two Japanese devils that gave us candies, but I don't remember who they were, they all looked the same.* They do not remember, but instead, they will remember all the bombardments, all the fathers and brothers who were arrested for no reason, sometimes disappearing, sometimes coming home crippled from the tortures, and all the images of rapes and hassling and the aftermaths of the war done at the hands of the Japanese Imperial soldiers.

End of March 1943: Life on Palau

Where did these women come from? Shirō wonders. How did these Japanese women, dressed in colors from before the war, red, yellow, orange, as if they know nothing of the state of emergency, end up on this small island in the south pacific, an island of Palau? Women walking around with parasols; women chattering like birds. Sleeping in houses formerly occupied by the

Europeans, sitting on the verandas fanning themselves during the day. They smell of home, though they are not like Masa; not at all. And he banishes that thought and keeps his mind to what is in front of him. They had landed only a week ago, but as soon as they embarked, they pitched their tents on the beach. How they ran around like little boys, shedding their summer combat fatigues and jumping into the sea. How they ate pineapple after pineapple, the first time they had eaten it like that, from trees, peeling the hard skin with knives, because for most of them, they had only dreamt of eating it from cans. And how they suffered from the numb tongues, from the diarrhea, and how they laughed. But that's where the boys' storybook ended. Then several days of going back to the anchored ships, practicing jumping from the ships in case of torpedoes, going up the ladder, then jumping into the sea with floats around their necks, with three dried bonitos and a water canteen hanging from their belts. A thought: *This isn't where we will be stationed, they are taking us somewhere across the sea to a different island.* Then they were taken into the jungle, through the jungle with all their gear on their backs in the sweltering heat, and after days and days of marching, they were given a tool each: now, build a shelter for your platoon. They cut down the trees, they cut open the path, and each platoon builds a hut. An elderly carpenter — or at least that's what he was before he was drafted — led the construction: find thick trees for pillars; find small twigs for the floor boards; find dry banana leaves for the thatch. It took them only a couple of days to erect this hut; it only took them a night sleeping under it to realize that it did not keep the mosquitos away, it did not keep away the rain that came thundering down, then disappeared, only to leave behind humidity. That they could not sleep well. This is what it means to be on a southern island: discomfort. This is what it means to sleep in the elements, in a way that Shirō has never experienced.

Sinking Ships

Ships carrying the red cross symbols make their way north, confident that they will not be attacked, sure of its universal sign. Yet. The Enemy submarines sneak up and fire torpedoes; the ships do not see them. It is night, and everything is quiet except for that moment when the torpedo hits the side of the ship. The steel body cracks open; the siren goes off in the ship. Comfort women, Japanese and Korean and Chinese, stacked in the cargo just like the way they came, labeled as *military supplies,* are locked in, and when the ships begin to sink, when the sailors begin to jump into the ocean one by three by dozens, swimming away as quickly as they can to get away from the whirlpool, when the wounded and the crazed soldiers are hoisted into the lifeboats, the women bang on the steel hatch, banging and screaming to be let out, but no one lets them out; they aren't supposed to be on the ship anyway. It is quick. The ship begins to sink. Letters, pages and pages of letters written by men scattered on the southern islands, all carrying the messages of their wellbeing, scatter on the surface of the sea as they, too, are pulled down by the whirlpool, down to the bottom of the ocean.

April 26, 1943: Shirō Leaving Palau

It is near midnight. No light. No moon. In the India ink blackness of the jungle by the sea, he waits with other men, holding their breaths as one, waiting for the whispered signal: *Now go.* They hoist up their gear onto their backs as one; they raise their rifles, as one, as they wade into the sea, onto the *Daihatsu* moored nearby that will take them to the ship anchored in the bay. In the *Daihatsu* in front of his unit, horses, horses, panicking and trying to kick their handlers. They have not been told where they are heading, only told to sleep with their floats as pillows, and in case of torpedoes, to jump into the sea and let their loincloths trail behind them against sharks. So they know, as one, that their Robinson Crusoe moment is done, that they are leaving behind a month of heaven to something malicious, something omniscient, like the *Daihatsu* heading toward the darkness and the unknowing.

May 1, 1943: The Coast of New Guinea

The sweltering heat in the cargo hold; horses dying from dehydration, even for such a short journey across a small sea; Engineer corps, tense, yelling, always on the lookout for an enemy plane, for one, because if there is one, there are more to come, like the angry buzzing sound; rumors, *heard that back in March, a convoy heading to New Guinea was attacked; almost no one survived; they all jumped out of ships, only to be gunned down by the Yanks; no, I heard the sharks got them.* A five-day journey from one island to another. Five days of fear, five days of small ships without defense, zigzagging through the sea; no planes to protect them, and the ships, the former postal ships and liners, follow the unfamiliar maneuver of the military movement, carry men as vulnerable as they are.

May 21, 1943: Death of the Great Commander in Chief of the Combined Imperial Fleet, Admiral Isoroku Yamamoto

The Central Headquarters did not release the news of his plane being shot down over the sky of Bougainville over a month ago. It did not say anything about how his plane slammed into the jungle, his body found strapped in his seat outside the plane, and an army doctor might have crawled over toward him before he, too, died. They made an announcement that he died gallantly in air combat, weaving another myth to keep the spirit of the war, and boys all vow to avenge the great admiral. And though they have only two years left, they will think that, because of his death, the war has taken on more mythical proportions.

May–June 1943: Working on a Runway at Wewak

As soon as they arrived at the bay of Wewak and the ships dropped anchor, they disembarked from their seafaring home of five days with all their gear on their backs, waded through the emerald water that came up to their shoulders, then to their waists, then to their ankles, panting hard as they dropped off their gear, and waded, went back to the ships to carry boxes of food, ammunition, disassembled artillery on their shoulders, passing the boxes from one person to another like a line of ants, leading the horses that panicked and struggled. That went on for hours and hours under the brutal sun, then suddenly the monsoon that fell hard on them like bullets, then stopped as quickly as it started, while standing in the sea, one box passing over their shoulder to another. The night came and they slept on the beach, out in the open, and as soon as the sun rose, they were told to pick up a shovel, a pick, and if there aren't any, to improvise: cut down the thick trees, pull out the roots, move the trees away, dig the earth, flatten the earth. In other words, construct an airstrip as quickly as they can, all by hand, all by labor. And under the southern island sea, they work and work until late afternoon when the monsoon came daily, then back to the site once again, day after day. But to Shirō, this almost feels familiar, almost like working on the farm, almost like being on the farm, tilling the earth.

May 29, 1943: The Last Telegraph from Attu Island

When the telegraph came from the Central Northern Territory Commander to the Attu Island Defense Commander Colonel Yasuyo Yamaguchi, he knew what it meant: abandoned. Holding the message of *The Army, with the help of the Navy, will do our best to rescue your regiment but do the best to fight against the enemy on your own. If all fails, it is our hope that you are ready to honor the name of the Imperial Army of His Majesty by Gyokusai — to fight until the last man,* he knew that he and his men have been left to die. It did not matter that they had wintered and barely survived for a year. Nothing mattered, except that the island has been abandoned. So he writes, *We have been attacked by the enemy on land, from the sea and sky, and the frontline battalions have nearly been obliterated. We are barely surviving each day. The lightly wounded and sick in the field hospitals have been asked to commit suicide there, and the doctors took care of the gravely wounded. Noncombatants are taking up arms to live and die as combatants. The Army and Navy have organized one company and will follow the attack unit. We are resolute not to live the life of shame as prisoners of war. Since there is no other choice, we will not dishonor the name of warrior at the last moments of our lives. We make our last charge as one.* And, after burning papers lest they be captured, he asks his communication officer to type in the fateful words: *Completed burning all the papers. I will now destroy the radio.* He commands that the sick and wounded must be disposed of. Without enough artillery, without enough ammunition, Colonel Yamazaki leads his remaining men, armed with swords and bayonets and charges.[4] That's how they find him several years after the war, his body at the forefront of other bodies. He led his men to death because that was what a soldier did, because he was commanded to do so, because the rescue was never going to come and men were dying all around him, because there was no other choice.

June 1, 1943: The Morphine and a Doctor on Attu Island

When the American soldiers burst into the field hospital in the basement, they find eighteen Japanese soldiers, nine on one side, nine on the other side, all tucked in under the blankets as if they are asleep, all with their hands in prayer. They did not suffer; a

---

4 "Attsu-to Gyokusai no Higeki: Yamasaki Yasuyo to Higuchi Kiichiro"「アッツ島玉砕の悲劇〜山崎保代と樋口季一郎」[Tragedies of the Battle of Attu and *Gyokusai*: Yamasaki Yasuyo and Higuchi Kiichiro], *Web Rekishi Kaido,* May 29, 2017, https://rekishikaido.php.co.jp/detail/3934.

morphine shot each was all it took to assist them to death. And in the alcove in the back, an Army doctor lay there, peacefully, with his head shot off.

June 1943: The American Way of Burying the Dead — Attu Island

The ground is still frozen, though it is early June. Summer does not arrive to this northern island, and when it does, it is as brief as spring. The bulldozers have dug a seven-feet-deep hole, a perfect geometry in this wild untamed island. They have scraped eight one-foot-deep grooves at the bottom of the hole to place a fallen soldier each, eight in total for each grave. But first. The war dead must be identified, and a medical officer goes from one body after another, disrobing each body to record the cause of death: types of wounds, number of wounds. Then he robes the body, moving on to the next one, only to do the same. Already men from the grave registration company have gone through the pockets, storing the personal items into a clean army-issued sock — always clean, never used — so that they can be sent back to headquarters first, then back to their families who do not know about the death yet. There can be no mistakes here. Every dead must be identified, accounted for. They take three sets of fingerprints. And now, each body is wrapped in a khaki blanket; they tie the body once around the neck, once around the waist, and once around the ankles. And they are gently placed into the hole. There are 125 bodies waiting to be buried today.

1943: American Soldiers versus Japanese Soldiers

The enemy stands 172.8 cm tall and weighs 68.38 kg. Our men stand 161.2 cm tall and weigh 52.3 kg. The average age of the enemy, twenty-six years old, and ours, twenty-seven years old. The enemy carries submachine guns and semi-automatic rifles; our men bolt-action rifles that must be reloaded every time a shot is fired. There is a foot of difference between them. And though they speak in different tongues, they speak of the same things: of home, of fear, of going home.

June 1943: A Rumor

Masa does not know who she heard it from, maybe she heard it from her brother, Gorō, or maybe Saburō heard it from one of his connections in the village office: the ships carrying the 51st, the one that Susumu may be on, sank on their way to New Guinea. Whom did she hear it from, she wonders, but then, so many times, she heard about this or that unit, this or that regiment was in China, in Burma, in the Philippines, when they were still stationed in Japan; about this man dying, only to find out that he is still alive. The radio plays the grave *Umi Yukaba* as it proclaims this or that battle won, but with sacred sacrifices,

> When I go to the sea
> My body will float amongst the waves
> When I go to the mountains
> My body will be amongst the grass
> For my lord, I will give up my life for free,
> For my lord, I will gladly give up my life.

And every time the song comes on from the radio, her heart stops, her ear strains to listen to which unit, which regiment, where, and every time she finds out it is not Shirō's unit, she sighs in relief. The rumors of casualty; news of winning. Which to trust, she is no longer sure.

A Soldier

When the self — or what we define as self — is stripped, one by one like layers of an onion, first status, then property, name, family, history, dreams, and desires, the only thing left to do is to find a new adornment, an easy self that one can wear like a uniform, and, taking on the role, *become* what the uniform represents. We are a representation, not the individual. We are anonymous, and in the faceless mass, we can do anything without worrying about the consequences.

1943: A Letter from Shirō to His Former Student, Kōji Maejima, from New Guinea

To Mr. Kōji Maejima,

Thank you for your letter. Congratulations on winning the Shooting and Sword Competition. I am very proud of you — the young man protecting the homeland. I am, as always, doing well. Jungle and humidity and monsoon rains are nothing. Compared to other southern islands, I was expecting tropical fruits, but instead, here, there are many coconuts filled with much milk. Enemy planes come every day and drop bombs on us. It's nothing but....

The Story of a Person

The story of a person is like any other. The story of a nation is like any other. We are the stories, we are the characters and storytellers, but at the same time we are not. Words of the war become the words of men and women, the rhetoric the same until there is no new story to tell except for the details. In America. In Japan. In Israel, Vietnam, Sarajevo, the Soviet Union, Sierre Leone. There is a boy — 18, 20, 23, it doesn't matter — who goes to war and he does not know what war is. He grew up listening to the stories of veterans who glorified their experiences, or who did not say much, either way he does not know what war is. His first battle, his first kill, his first personal loss, that moment when he loses his innocence, that moment of clarity when he stares into the eyes of the enemy or horses and he became the character of the old, the new, the killer, the warrior, but his heart is numb and his soul, somewhere else. And if he dies in the war, someone will take up his story, someone will beatify him, *see how he sacrificed his life for the nation, for freedom, for liberty, for our motherland, for...* and if he lives through the war, if he makes it home, he carries the war within him without words to talk about it. He will not be the boy who left home only a handful of years ago, and that time, the time in war, will mold him, mark him, give meaning to his life after. This is all easy to say but remember: there is a person to every soldier. The details. The lives interrupted; lives fractured. Their first kisses, their first kill, their heartaches. There is more to the story, but there is no word; sometimes, words fail them.

August 16 and 17, 1943: The Bombing of Wewak

It came as a droning of mosquitoes, or was it flies, and Shirō waves them off, as it has gotten to be the habit of all men working in the airfield, waving their hands to slap off the insects as they stomp on the earth, as they cut down the trees, as they carry the limbs. It came as a droning so far away that at first Shirō looks at the sky to see whether the monsoon clouds are rolling in, the weighty sky, but the sky is clear and mosquitoes and flies still buzzing. Then the sudden air raid siren: the western sky covered with bombers, closing in, closer, closing in: men drop whatever tools they have in hand and scurry for cover; men push each other into the bomb shelters, into the jungle; *take cover, take cover.* Shirō rolls into a shelter after his men, *down, down,* shouts and yells; the metallic shrills, then the earth starts to explode, here, there, everywhere, covering every inch of the earth; men start to dance amidst the hailstorm of bullets, blood spurting, shooting out; men jump, fall, under the strafing, and Shirō curls up in the ball, just like others in the fox hole, *oh shit, oh shit, oh shit,* chants begin, and beastly screams drone under the zooming planes, and bodies begin to rock with the chant, and they are helpless, these men on the ground, as helpless as ants being drowned by boys watering their hole for sport, for fun; the moans, the screams, *someone help me, I've been hit, help me, help me,* the bullets screaming out, the earth changing under the bullets, men are pissing in their pants, the earth turning bright red with fresh blood and exploded body parts and orphaned limbs as the planes carrying the insignia of the rising sun fly away from the airfield, from the enemy, from their own, to Hollandia, leaving the men defenseless.

September 1943: The Lucky One

Susumu was the lucky one. If he had been one of the 873 men to survive the trip to Lae, New Guinea, instead of one of the 4,543 dead men, he would have discovered that Lae did not offer comfort or rest but survival. The 51st Division, 18th Army is surrounded by Australians. They have nowhere to go except to confront the enemy with full force, until the last man, but the command comes from the 18th Army general: find a way to Kiria, or Madan, no matter what. There is only one way, by going over the Salawaket, a mere 120 km between Kiria and Lae. A former Olympic marathon runner is chosen to find a path to climb over the 4,100-meter mountain, and he and the New Guinean guides make it in twenty-two days. It is decided. This is the path to mobilize — or is it "retreat" — 8,500 men. The calculations are done: a man can carry ten days' worth of supplies and no more, but if they can walk 16 km a day, then they can make it to Kiria in sixteen days. If they are careful with their supplies, then.... The field artillery of the 14th Regiment, Susumu's, decides that it will take one artillery, at least, but abandons it quickly. Who would have known that after the thickest jungle, there awaited them an open plain that went on and on, making them such easy targets for the enemy that they can move around only at night, and if they survived that, the narrowest path across a mountain pass. The temperature goes down to below freezing as men already weak from hunger, from malaria, misstep and fall down the ravine. One by one, men fall behind, *Go ahead,*

I'll catch up with you, and no one waits for them. They can care only about themselves. They can carry only their own burden, and make their exhausted bodies go forward. Compassion has no place when there is no luxury of survival. Keep walking, one step at a time. One out of four men does not make this trek. Susumu would not have made it alive. Or maybe he might have.

September 10, 1943: The Quiet Radio

There, in Tottori prefecture facing the Sea of Japan, an earthquake kills 1,083. There is no announcement on the radio.

September 13, 1943: Flag Ceremony

The sacred regimental flag, the one guarded by the flag corps, the one that was blessed and handed to the regimental commander by His Majesty the God himself, the embodiment of the regimental spirit itself: today is its birthday. After the tense air raids of having lost the airstrip, after having to rebuild an airstrip in two days, after working on it, by hand, by their breaking backs, for three months, this comes as a relief: a day of not working. A day of celebrating, when the tightly locked food storage is opened up, when they say that the alcohol will be freely distributed, that men will eat and dance. For now, the regimental ceremony is a solemn affair, where they raise the flag in the northern corner of the base, and they all bow deeply northward, where the Emperor–god and his family reside. A speech. Then another. About the holy war, about the sacred mission of the Japanese Imperial Army, about the Japanese spirit that can conquer any foe, any trials, just by sheer will. Then, in the late afternoon, men relax their stances, they divide up in companies and units, and someone starts to sing a song from home, another starts to dance, and they drink and eat and they are relaxed. A day of smoking, one cigarette after another, not caring about preserving the limited supplies, warding off the flies and mosquitoes, because today is a celebration, because today, life is easier, the day they can forget about aerial bombings, of the enemy that is stronger than them. They sing together, as one, some men pretend to dance the dance of the natives, and laugh and talk of home and boast and tell tall tales, and officers join in, and it is almost like it was back in China for Shirō, where it is a brotherhood of men, no longer tense, no longer tired and suspicious, but enjoying the moment of peace, just a moment, but enough for now.

October 1943: Shirō Writing Letters

There hasn't been any mail drop-off from the planes and trucks as they were used to in China; there haven't been horses and donkeys carrying the familiar care packages and letters — for horses die like flies here on the southern island and no natives have cattle to do their work. On days he is working, on days he is not leading the drills, at night under the kerosene light inside of the mosquito net, he writes in his journals, as he has been doing for the past seven years, a habit so ingrained that he writes as a sergeant is supposed to write, to keep an objective record of his role as a leader; he writes letters to forget about the mosquitos that somehow make their ways inside and buzz around his ears; he writes to forget about the jungle right behind his shack where at night strange wails come out, ghostly wails of birds, of animals, or maybe men sick with malaria or jungle madness; he writes not of the jungle he does not understand, of the surprise air raids by the enemy, where they follow the trails of smoke floating out of the canopy of trees; he writes not of the limited food he eats out of cans that taste of rust; he writes of the island of his imagination, of the happier place full of coconuts and flying birds the color of the rainbow, he writes of the emerald sea and the coral reefs, just like he had read as a boy growing up, and he writes of his memory back home, of his wife, his three children, but he does not write of homesickness so keen he sometimes wakes up in near tears that he has to wipe off before his men see him.

October 1943: Shirō, Wewak, Papua New Guinea

Too many kinds of death can happen on a battlefield, and New Guinea redefines battle for many soldiers, even the veterans like Shirō, who has seen so many different ways to die, so many different ways for a man to get damaged. China was a country that was still familiar, with the same faces as the people they had left behind, with lands cultivated so that Japanese soldiers could pillage, steal, and even rape without hesitation. It was still a cultivated land with a history that went back several millennia. Here in Papua New Guinea, where dark-skinned men still live in huts, where women walk around with their breasts bare and men with spears in their hands, the land itself is wild. This is what Shirō sees when he first sees the coast of New Guinea. He is bewildered by the thick, green landscape, by the men who speak different languages according to their tribes — even though they are separated only by a mangrove — with Pidgin English as their common language. This is where fire and water, these elements that civilization has appropriated and controlled, are as rare as food and shelter. Nature is cruel and indiscriminate here, and Shirō's soldiers, with their technology, and the Emperor, with his expanding empire, do not mean anything. Guns rot in weeks; artilleries rust soon enough. Tanks have no destructive power when they cannot find petrol or even roads. Horses wilt and die of starvation. Even the Emperor's soldiers must bow down to the elements.

November 1943: Papua New Guinea

They rename the landscape, one place after another, undoing the former colonizer's names, and they rename one river after another into the familiar language: Driniumor river becomes

Bandō river, and here and there, Tazaki, Tsurumaki, are renamed after the rivers back in the *innerland,* to conjure up the memory of their homes they can no longer remember, not even when they close their eyes at night. To ease the disorientation, the homesickness, they tame the unfamiliar jungle island into a familiar place, a place they can, for a moment, mistake for what they left behind six thousand miles north.

The Rule of the Jungle

Do not walk alone; always walk at least in a pair; do not look down as you walk; do not focus your eyes on what is immediately in front of you, but focus on the opening, the break in the dense growth, through the growth, not at the growth; do not rely on a map — the map is not detailed and is nearly forty years old; always carry water with you, and make sure you have enough to last for the journey; do not rely on stream water; be sure not to use your strength needlessly, do not cut the growth needlessly, use sticks to push away the undergrowth; make sure you leave some markings on the branches, on the path, just in case you lose your way; carry a fistful of salt, just in case you lose too much when sweating; follow the beast trails, they always lead to water; if you are crossing the river and it begins to rain, move as quickly as you can out of the water or else the sudden current will carry you away; if you need to drink out of the river, dig a meter away from the current itself, the clear water will seep into the hole; be sure to carry live fire to build a big fire with, for matches rot with humidity in the jungle; stop and listen to the sound in the jungle often — sound does not travel the way it does on beaches or forests; do not be scared of bird calls which sound like babies wailing; do not touch insects and snakes needlessly; do not try to grab vines needlessly — it may be poisonous or thorned; do not eat things in the jungle; do not shoot your gun and rifles, the bullets can ricochet; do not use your sword to cut the undergrowth, your sword is the spirit of the warrior; never abandon your platoon and regiment; never travel through the jungle unauthorized; when doing laundry, never dry your clothes or build a fire on the riverbank, or you are announcing your presence to the target; the tree canopy shuts out the light inside the jungle, do not rely on the sun or the moon for direction; do not travel at night; do not travel alone. This is a mysterious, malevolent place that rejects men, which even the natives call *my kai kai jungle* — man-eating jungle.

Undated, 1943: The Last Communication — A Postcard from Shirō to His Family Upon Hearing of Susumu's Death

I'm sorry for the long silence. I can't hide the shock and grief over the news of Susumu's death. Please make sure that his memory is not tainted at the memorial, and honor his soul as I would have done if I were there. I did not do enough for my young brother while he was alive, and, being so far away, I cannot take care of him even in his death. I am in deep sorrow — please understand how I feel. My parents' grief must be beyond words. Please look after them for me. Do not dishonor him in his death. Look after the children. Take care of yourself, promise me that.

Dying Alone, Dying Together

Who is to say dying in battles together is foolish? Who is to say that when one has to die, one wants to die with one's brothers, more than brothers, the bond so strong even blood is repelled by the love? Who is to say dying together is silly when all we have done together, shitting together, eating together, telling each other stories of the life before all this, watching others die and crying together, getting shot together and thanking gods that you survived another day, freezing together and holding each other not as lovers do but as brothers because we are cold and we could die if we didn't? It is better than dying alone, or living alone in a jungle, although we know that we all must die

THE WAR WITHOUT AN END

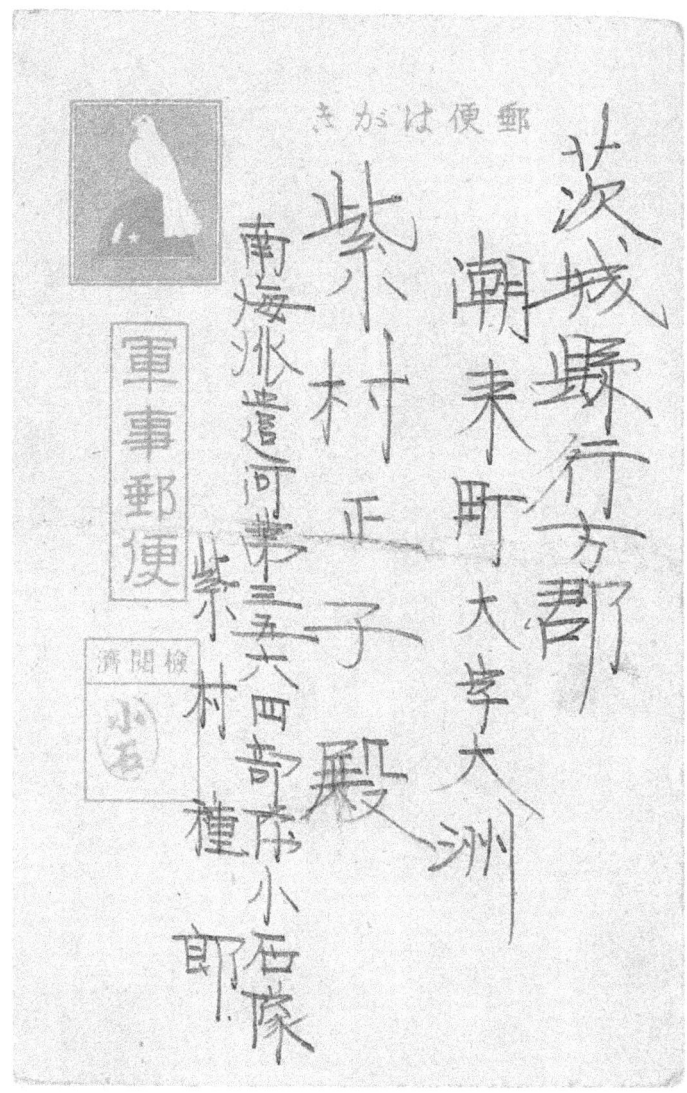

Fig. 5.3. The last communication from Shirō, front.

IMAGINARY DEATH

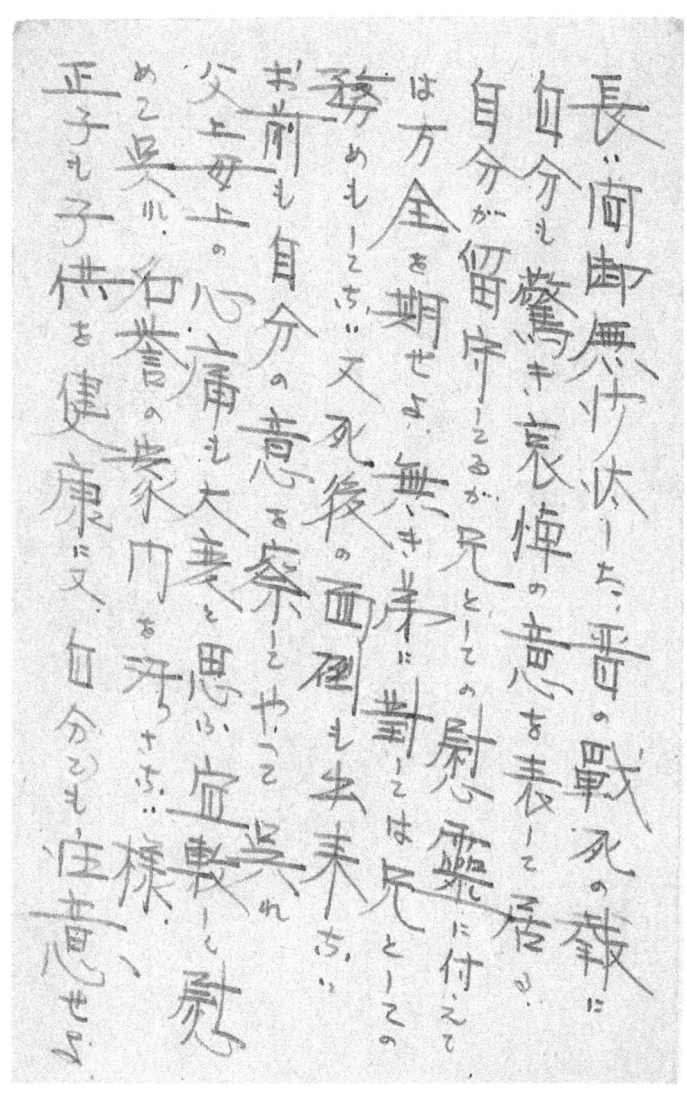

Fig. 5.4. The last communication from Shirō, back.

alone, each in each, without fanfare, without celebration, without vigil and mourning.

December 1943: The Final Documented Contact

On a translucent piece of rice paper, the final communication from Shirō: twenty *yen,* sent via postal savings, to Masa. His battalion name. His rank. But the paper is so thin, and the words barely visible, probably arriving with one of the last ships that made it safely to a Japanese port. With his body and life, he exchanges it for twenty *yen.*

December 1943: There Is Democracy in War

Women stand in lines with their ration coupons, even the baronesses and duchesses, only to find the shelves empty and nothing to eat. They stand with aprons, just like anyone else, though theirs are better quality, and build engines for airplanes, pack care packages. Everyone lives in darkness, the windows blackened and fearing the rumor of the enemy air raids. Men, too old to be drafted, are taken away, in bulk. Even an assistant professor of political philosophy at the Imperial University, Masao Maruyama, is not immune. At age thirty, he is drafted into the army as a private, and spends his basic training being slapped around by the illiterate, by the uneducated, and later he will be in Hiroshima the day the bomb falls from the sky, turning the city into a ball of light for an instant, the light that some saw on the other side of the ocean in China. He will stand in front of students in the 1960s, attacked, lynched, for being the symbol of post-war democracy. But that is in the future; right now, on this day, he is drafted. So are men of nineteen, and men as old as forty-five. The age range has been stretched. Now, everyone is equal. Everyone must shoulder the war, embrace it with their arms, with their entire body, if not with their heart.

IMAGINARY DEATH

December 1, 1943: The Start of the Mobilization

The journey, they tell him, will be 300 km, or to be exact, 303.41 km, if in a straight line, but the coast path from Wewak to Madang is not straight, is not an easy path along the sea. On average, on a flat ground, a company can travel 24 km a day, a journey this long will take 12.5 days. But Shirō's officer tells him that it will be a night march, traveling in platoons, so slightly faster. They are given a week to Madang. They will always walk toward the sun, travel first along the coastal path, then into the jungle, through the jungle, across two rivers, then through the jungle again, then along the coastal path, to Madang. And if one brother is getting ready to march across the land, another brother, the youngest, Kōkichi, is traveling on a boat across the coasts of Asia, from Busan to Taiwan to Manila, never staying in one port long enough to acquire a new language, new stories, or news of his brothers, one already dead and another in New Guinea.

December 1943: The Collective Dreams of Men

Men all dream the same dream: of waking up one morning, and instead of waking up under the damp, exacting jungle canopy, where their bodies sink low to the ground, they instead wake up in their beds, on the familiar straw mat of their home, waking up to the voices of their children. And their wives turning around, *You're up; breakfast is ready.* They stretch, they turn their bodies, smelling the pillow that smells of recently washed hair, *theirs,* smelling the straw mat, smelling the faint sweet smell of rice they have not eaten in such a long time. Their wives walk toward them, their soles making a faint sucking sound against the dry floor; they lean down, and whisper, *Wake up, it's time to wake up.* The aroma of an unbathed woman, still smelling of sleep, the faint smell of cooked rice, the green summer smell of cucumber, the freshly cut smell of haystack, the night soil, the earth, and underneath it all, the musk of working on the land.

Of homesickness. And that is always when they wake up. They sleep next to a dead body; they sleep half buried in the ground of thousands and years of rotten leaves, of rain that never warms them, of mosquitoes the size of a fly and flies that land on the unblinking eyes of the dying. They sleep in exhaustion, in fear of dying.

The Malicious Jungle

The jungle keeps giving. Already, men have been given the gift of the jungle: malaria, dysentery, beriberi, perpetrated by extreme exhaustion and malnutrition, your feet are never dry as you walk in the jungle, wet vegetation and mud holding onto your ankles, it makes you feel just like back home, when your children held on to your legs in jest and you laughed as you took steps, one at a time, but this is not a laughing matter anymore. And shoes, after a month in the jungle, rotten off, and once you take off your shoes, your feet look as if you are suffering from athlete's foot, but worse, covered in blisters and skin coming off in big patches. But it does not matter if you are suffering from jungle feet, whether you are suffering from malaria where fever flashes through you from head to toe, where your ears ring so much you can't hear anyone's voices, not the mortar sound, not the bullets, whether you are suffering from dysentery so much that you have slept with your pants around the ankles because you already have malaria so you are no longer afraid of mosquitoes, because when you move, nothing comes out except for whitish liquid and flies have laid eggs on your anus: when planes start strafing you run, run, run, no matter what your condition, and if you don't care, you have already given up on your life.

The Mobilization Order from Wewak to Madang

Walking along the emerald sea to the left, the jungle to the right, then after a night of walking with their all their gear on their backs, they retreated into the thicket to rest in the dark jungle where the sun doesn't enter. They fell with their backpacks on their backs and fell asleep, just like that. Then as the sun set, the command to wake up and start marching, marching through the beast trail with twisting vines that hook the feet, the jungle full of wails of night birds and unseen ghosts, and they walk, holding on to each other's backpack belts, one line after another, as someone jokes, *I heard elephants do this when they walk.* They do not say that they are scared, that they don't understand the jungle; they pretend to be brave, though their steps are furtive and unsure. The jungle: arrested time. Light suspended above the canopy of trees; the moon does not exist in this landscape. Shirō walks ahead, in front the platoon commander, pushing away the vines, whispering, *Watch out, there's a thorny vine here; watch it, there's a branch at the head level,* whispers traveling from one man to another. Four days of marching through the night, a sight: a glowing tree in the middle of the jungle, ghostly, majestically. A tree covered with lightning bugs. A tree. And the march stops. All men gasp as one, their feet stop, and for a minute, they forget that they are in the jungle, that they are in a war, a holy war, that they are fighting for the Emperor–god; for a second, they each remember their own lightning bugs from their own neighborhoods in Ibaraki, back home, home where they came from. And after a night of marching: the river.

1943: How to Tell If Someone Is About to Die: A Saying amongst Men in Guadalcanal

If you can stand up, you have thirty days.
If you can sit up, you have three weeks.
If you can't sit up, you'll have a week.
If you can't get up to use the bathroom, you'll have three days.
If you've stopped talking, you have two days.
If you've stopped blinking, you will die tomorrow.

January 1944: No New Year

As Kōkichi is hopping from one island to another, this time from Manila to Cebu to Negros island to Malay, Shirō is still walking, this time away from Madang, this time to a village called Onishi east of Madang. You will know, he was told, so he stood by his officer as the officer told their tired men, *Okay, we need to march again,* and how they sighed and grumbled, but they were beyond tired, beyond complaining, they all got up like ghosts, hoisted up their dwindling supplies, they no longer slap off mosquitos because they have gotten so used to them, they got so used to the insistent buzzing and any exposed flesh eaten alive, or so it felt like, by mosquitoes, and they readjusted their shoes they have not taken off for such a long time, the bottom so thinned down that they could feel every nook, every stone on the beach, and they begin to walk, one step at a time. It is New Year's, but no one remembers that; no one remembers to bow deeply to the north, where the holy Emperor is, where their homes are.

January 1944: In the Battle, There Is No Logic, in Death, in Living

Who lives. Who dies. There is no logic to it. A small, slight wound turns into a monstrous gape in the flesh overnight. A cowardly man survives while the most hopeful one, who believes in his own immortality, dies so quickly and insignificantly that his own death astonishes him. A man can be walking behind you one minute, only to be gone the next without an utterance, as if swallowed up by the jungle. The toughest of men can succumb to malaria, but the malaria does not kill. It weakens. The softest, the weakest, take shallow breaths, their prayers hovering around their tongues, and the gods hear their prayers. Men are left behind in shacks when they cannot walk on their own; men are left behind on the jungle path, still alive, while their brother-soldiers walk away. The jungle is full of voices moaning, *Don't leave us here, please don't leave us here.* Men die from honesty, from hunger so keen their bodies transform into the grotesque, from malaria, from infection, from hope, from bullets and mortars from the invisible enemy. Men gorge on love, on hope, on the gods, on memories of home and their helplessness. Men die. They live. Fate stalks the men on this island battlefield. No one can barter with the gods here.

January 1944: New Guinea

Americans and Australians are no longer the enemy. Those enemies are the luxury of generals far away in the innerland who can afford to wage war on the nation. The enemy is no longer the American devils. And enemy planes that drop tons of bombs during the day — almost never during the night — are irrelevant compared to what is in front of him. Here in New Guinea, the enemy is visible, but there's no way to combat him because he lives within, gorging on bodies from the inside: the enemy is hunger, insanity, fear. And it is not hopelessness that kills a man. Death is indiscriminate: men with hope, with the will to live, die; men without hope also die. In the truest sense, class or rank ceases to matter in the jungle. All that matters is that, with the supply line completely cut off, New Guinea has become an isolated island, where there is no food, no cultivated land, no logic or familiarity. This is a land that denies life to the intruders.

1944: Mae Doe, or the Big Death

The New Guinean tribesmen don't understand it. Yellow and white men came from the ocean in steel boats one day. They came with their firing sticks and rumbling metal beasts, they spoke the tongues of beasts. What is there to hate these strangers, the tribesmen wondered. In our tribes, we fight because we hate, because we are angry because they stole our fruit, our women, our children. There is no other reason. But these men.

They come and cut down trees; they flatten the land and destroy our food. They do not stop after one or two people are wounded. They keep fighting, they keep shooting at each other. There is no hate between them. The tribesmen call this *the Big Death*; it is death so big they do not have the concept of a number to describe all these deaths.

February 1944: A Notebook of the Dead

To be shot. To die in battle, that is the only way for a soldier to die. But Shirō also knows that only a handful die a clean death: a shot to the heart or the head while charging. Most die waiting: from malaria, from strafing, bombing, starvation, insanity. It is his job to keep track of the dead. He writes down all the names of the dead in his squad; next to it, he notes the times, the dates, and places. His watch has been broken for a while, so he guesses the time. Under "cause of death," he writes, *death in battle*. Always. It is the small comfort he can offer the survivors, the people back home he's heard so much about on this journey; he's heard so much about them that they feel like distant relatives he's never met. To honor his men, to ease the pain of the fallen, he writes, *death in battle*. No one wants to hear that her son or husband died of malaria, a lingering, maddening disease that ate up mind and body. Or that a man shot himself in the foot by accident, and the wound festered, with the sound of maggots eating the flesh so loud at night. How this man could no longer walk, and there was no doctor to amputate his leg, and no one was strong enough to help him walk. They themselves weren't strong enough to walk without crutches. So they left him with a hand grenade. And he accepted. No one looked him in the eye when they said goodbye. He kept crying, sobbing, but he did not say a word, as if he understood. Another man disappeared without a sound while crossing a river. *All of them died in battle,* he tells himself. And if and when he dies, the next-in-command will take the notebook. That's how the notebook came to be his responsibility in the first place. And the notebook, even after the last man, will keep going. Names and dates written down. Places unknown, because the writers didn't know where they were. Names written by different hands, the transition from one hand to another seamless, but understood: the writer had died. The next writer wrote down the name of the writer before him, as Shirō does now. The next writer will write down Shirō's name. His wife will know he died — without details. And his children will not remember him as a man, but the notebook will go on. He will be remembered as a soldier. He will die as one. As an Imperial soldier.

1944: Humidity Is the New Season

There is no winter here; there is no spring. Only humidity. And salt. No one ever said that men need salt more than love, more than sex, that water is needed, but salt as well. The last letter he had sent was the receipt for his pay but there's been an uneasy joking amongst them, *Heard all the ships are being sunk by torpedoes,* and he wonders whether the letters he sent home have gotten there safely. Because he hasn't gotten any letters in so many months, and he has read the letters he was sent again and again, remembering all the words, but he now fears opening them anymore. The papers can't stand this humidity, and all the photos are curling around the edges, fraying and ripping, especially Masa's face where he has caressed so many times that right in the middle of her beautiful face is a fold. He closes his eyes and sometimes he can see her, he can almost imagine his children's babies so many years into the future, and sometimes, the faces he so loves, faces that have kept him going this far, are blurring, details as vague as something at the bottom of the river. When the malaria gets to him, he feels cold from his head to his toes, and then fever courses through his veins and he can't remember anything, he feels delirious. All the men have malaria in different degrees of affliction, and the only thing they can do is to lie as still as they can, because there is no medicine, because there is no one well enough to bring water, there is no doctor who does not have malaria, and because the food is rationed so much that they only eat once a day. More and more, his home, his memory, is slipping through him and how he is holding on to it, holding on, holding on before he will start living this life, without past, without memory, without *them*.

March 1944: There Is No Way In or Out

There is no boat out of here carrying a Red Cross flag for the safe journey back to the *innerland* — home, where his family is. There is no supply boat carrying reinforcements or the much-needed supplies or letters from home he yearns for so much that his heart aches. The supply line has been cut off; the only new men Shirō has seen have been merchant marines who had attempted to land supplies, only to get their ships sunk right in front of their eyes. Lives are cheap here. And he envies the enemy, who plans so carefully, who has more ammunition, more food, more men. He has seen canned food strewn about, has heard about a mess hall where a cook cooks warm food for all. When was the last time he ate anything warm? Four months ago? Was that when the orders had come not to light fires? When was the last time he closed his eyes and woke up rested, not fearful? Someone had said that the enemy allowed only 40% casualties in any given battle, and that when they said *casualty*, they meant the wounded, not the dead. *How different that is from us,* he thinks. Here, casualty meant sure death. He has heard of units charging in desperation toward the enemy with only swords in their hands. *Is life cheap? No,* he reasons. This seemingly desperate act was actually a hopeful act — an act that defied all meaning — but underneath it all, there was hope: to break the will of the enemy. To win a battle, only if there was only one man left. If the enemy retreated, the battle was won. And battles won may lead to winning the war. And after that, when the war is won, he can go home.

March 1944: The Breaking Point

After 200 to 240 days in combat, all men break down, even the strongest amongst them. For Shirō and his regimental unit, that would have been January 1944. With constant aerial bombardment, strafing, and the fighting with the unfamiliar elements and hunger and fear, this month would have been a breaking point, men shutting down completely, not being able to make a coherent decision, following orders without questioning because there is no more rational faculty left, the bodies reaching the breaking point that their minds leave while the bodies just lie there lethargically, sexual drive gone, their genitals receding as if to announce the departure of the need. This should have been a breaking point, but Shirō somehow kept going.

March 1944: Everything Rots

Swords rust. Guns, both big and small, are rendered useless when there are no bullets left. Planes without fuel become earthbound. Medics and doctors can do little to save lives when there is no morphine to ease the pain, when there are no surgical tools to amputate useless limbs. The most precious thing on this island is a fistful of salt that you carry with you. The next most precious thing is fire. And you save the last hand grenade for yourself, just in case. Here in New Guinea, whether you are a pilot, or in a field artillery, or in the Navy, you become an infantry man at the end, to march, to retreat, always on your feet, on your exhausted feet that have been walking non-stop for a year now.

March 1944: New Guinea

A thought crosses Shirō's mind: *Maybe it is the weak who know their limits. Maybe it is the weak who can sniff out death and run away from it as quickly as they can. Maybe the strong, too eager to prove themselves, too attached to the sense of honor and loyalty are the men who die first.* And he wonders how he has lasted as long as he has. And he wonders about his two younger brothers, all scattered somewhere in southern Asia. Susumu was on his way to New Guinea when his boat was sunk; that's what he had heard from a man who heard from another man who heard it from... well, that's how they get the news here, all through hearsay. *Susumu,* he thinks, and his heart darkens. Susumu never wanted to go to war, and the family thought that because Shirō had already served, Susumu wouldn't have to. But then the Japanese Navy attacked Pearl Harbor, declaring war on America. Susumu was not weak. He was strong yet there was nothing he could do when the boat was sunk by the enemy.... Shirō shakes his head. And Kōkichi. For some reason, Shirō knows that Kōkichi will live through the war because Kōkichi does not take anything seriously. He never did. Maybe it was the blessing of the third son, with no obligation to the family. Maybe that was why he never got good grades; he was smart, but not book smart. He had surprised everyone when he studied for only three days — a mere three days! — and got into the Teacher's College from the Agricultural School. No one from the Agricultural School got into Teacher's Colleges. Kōkichi, how he laughed. Maybe he will laugh his way through the war as he has always done. Light, carefree. Never the way Shirō was. Nor Susumu. Maybe Kōkichi will be the one to survive. And suddenly, a thought crosses Shirō's mind like a bird taking flight from a tree branch, startling him: *Am I a weak person for having survived so long? Is that why I have lived when stronger men have died so quickly?* He rolls that idea in his mouth because he is hungry, and he has not eaten fully for such a long time. And when his saliva has made that idea soft, chewable, digestible, he shakes his head. *No, I am not weak, but neither am I strong. I just want to go home.*

March 26, 1944: A New Order

They are given two weeks' supplies and march to Hansa, right between Madang and Wewak. Retracing the path they had taken only three months ago, when they arrived through Madang to Oishi Village a day's walk away. And they know two weeks' supplies won't keep them alive, not with the route they have taken. A sense of dread weighs them down; men around Shirō are sick with malaria, with weakness, with jungle fever; Shirō is sick as well. A command: leave the ones who can't walk in the field hospital. Ones who can, pack up and now, move; the ones who can move, those ones do, but do not know if they can make this trek again.

End of March 1944: Kōkichi at Sea

He never stays too long in one place, adrift at sea, almost like Hiruko, the leech child, the forgotten god-child who was put into a basket at age three because he had no legs and arms, because he was a product of a mistaken courtship ritual. Kōkichi and his unit are like the forgotten child, he thinks, only that at every port, every cove, there, always a command awaits him: go there, do this, and he does. It all seems like a useless command, pick up this man, pick up that man, do not deviate from the command. Why pick up only one person when there are so many at the beach, wounded and beyond fighting, to be transported home; why travel to this shore empty handed, only to pick up a light cargo? He does what he is commanded to do, from Saigon to Singapore to Penang, staying at each shore enough to enjoy the cuisines, enough to walk around in its glorious cities where men and women bow at him, but not long enough to see the darkened faces of these people. Letters in a water-proof package thrown into the sea, a bamboo used as a float for the tide to carry it to the beach; a package of food and ammunitions thrown into the sea, not enough to feed men, not enough for one man.

April 1944: The Enemy Comes Down Fully Equipped

When they come, they carry with them the comfort of home, luxury beyond our imagination. The airstrips that we created with our bare hands, trees felled by axes, one at a time, the earth dug out with small picks and shovels, then flattened out by stomping with our feet. An airstrip that is created one meter at a time, only to be washed away by rain that came every day, every other day. It took us eight months to create a 1,500-meter runway. When they come from their ships, they bring with them bulldozers, an airstrip emerging in one cycle of the moon. Those mosquitoes that broke down everyone in the unit, mosquitoes that rendered us useless as we shake, sweat, every three days, every week, those mosquitoes are destroyed even before the enemy's feet touch the ground. When the enemy aviators fall from the sky, they come down with the whitest parachutes that gleam, shine brilliant in the sun, and they carry with them a fishing hook, a rubber boat and red dye, food, a smoke pot, one for the day, and another for the night, and when an aviator falls, planes immediately come. Everyone is accounted for: missing, dead, alive. When the enemy comes, they throw live cows with parachutes from the planes. The enemy come with radios and warm food, with clean uniforms and a never-ending supply of ammunition; they come as if they are on a hunting trip, while we sit in the jungle, foraging for food, counting our ammunition, waiting for letters from home, for a ship that dared to cross the ocean to pick us up from this island, and we have become the prey, running, hiding, moving only at night.

April 1944: Staying in One Place

For the first time in four months, Shirō and his men are told to stay in one place, but there is no supply station anywhere; there is no clean water anywhere in the vicinity at their base in the camp, hidden from the enemy eye. Any semblance of discipline is ignored by the men's exhaustion; they cannot move, they are lethargic, they are tired. Everything they do is slow: carrying water, cutting down trees, moving, even mere walking. Shirō wonders about the men he had left behind in the field hospitals scattered between here and Wewak, one man here, two men there. Perhaps they are better fed than he is; perhaps they are put on the boat to Rabaul or Palau, then homeward on a Red Cross boat. Maybe they are in a better place, he thinks. It is a new language, new landscape they have to learn: what is edible, what is not; which bird is edible, which is not; how to cultivate taro roots; how to stay alive because there is no supply station, there is no canned food or ammunition anywhere at hand. They do not yet know that many in the military command do not want to admit that they have overexpanded, stretched too thin, that they do not think that men cannot feed themselves in the jungle, not like they did in China where they pillaged and looted and took away from peasants they were supposed to be liberating, like they did in Manila and Saigon and Hong Kong—the war supplied by the native inhabitants of the land.

May 1944: Rice, Hunger

He knows that back home, it is the rice planting season. They will wake the field that has rested during winter, they will turn the ground upside down to pluck out the roots that have taken root during the short burst of spring. They will wake the irrigation drawn from the river behind his home, which geometrically followed the pattern of the field; Masa will stand knee-deep in the mud, planting the rice shoots in a straight row. He himself would have done that if he were there himself. He wonders whether they have enough men, with Kōkichi gone, with him gone, all men between the ages of 19 to 40 years drafted to the war. How will they take care of the field, how will they... and he pushes the thought aside to come back to the present, to what is important, to the men lying around lethargically, having lost the will to fight the malaria because the body has no energy left, not with enough food left. The rationed food has already run out. He has not told his men: it's a secret between his officer and himself.

Letters, Photos, Wetness

He has been so wet for so long, *when was the last time he was dry?* Even when he hasn't been rained on, there's a persistent wetness — from sweat, from humidity. His past, slowly erased by the elements; his body, devoured by the unknown. Papers have run out a long time ago; photos are disintegrating in his pockets. As he waits under the mangrove forest for the monsoon to subside, he wonders whether he, too, would disappear as he has seen the bodies, disintegrating in the elements in a blink of the eyes, the insects carrying out their duties in half a day. That he would be anonymous in his death, as he has seen men's bodies carelessly thrown into the holes, as he has seen bodies, their uniforms barely intact to identify them as Japanese, huddling under the trees, on the path, vegetation overtaking them, swallowing them, making them a part of the jungle they so feared.

1944: A Father's Secret Wish

It is late, and he is tired, but Saburō cannot sleep. Sleep does not come easily to him anymore, not after his three sons have been *taken away* by the Emperor. *Taken away*, he thinks, then stops himself. No, they were not taken away, they were *given* to the Emperor, because the Emperor called for them. He called not just his sons, but boys and men of all ages now, not only the young. Saburō tells himself, *They were not mine to start out with, they belong to His Majesty the Emperor.* He listens carefully to Eiko sleeping next to him; her breathing is even, slowed, and he knows that his wife is asleep. He slowly raises his body from the warm bedding, and carefully steps over his sleeping children's bodies, dead asleep from the day's work on the farm. He knows the way, even in this darkness — this room, this house, the neighborhood so intimate that he can take himself through the dark path to the shrine of the water god. He has done this so many times, this begging, this bargaining with the god. This is not what a man is supposed to be doing, but no one witnesses this: an old man making a plea bargain with a god because there is only one god now. *Please protect my sons, please make sure they can return to me. I have been your faithful servant; please let them come back to me.* He does not see that on other nights, during other hours of the night, his neighbors come out to the very spot he is standing to ask for the same thing, even the most fanatic servants of the State who yell out, *Die honorably for the Emperor,* in the daylight.

The Will of the People, the Yamato Spirit

With this spirit, they tell us, we can do anything; we can work through the night and day in the factory, not stopping our hands to rest to make ammunition and tanks; with the Emperor-blessed spirit, we can conquer any disease, we can suppress our desires for more food, for more sleep, for loss, and we can kill the enemy with our bare hands, even a child. Schools are closed; students, those old enough not to be forced to leave home for evacuation into the countryside, work many hours in factories, a handful of them even working to mine uranium with their bare hands, although they do not know it yet. Magazines are shut down, even the ones that have followed the government censorship, because there is not enough paper around. Central Headquarters tells the regimental commanders, *Tell your men to find their own food; tell your men to fight without them.* Central Headquarters, that mysterious puppet-master, tells everyone, *Hunger is nothing, if you have time to think of food, you are being selfish; think of others, think of men fighting in jungles, in cities surrounded by the enemy on four sides.* And we drink their words and believe them, even though the darkest part of us knows better.

The Broken Ones

The first ones to break in battle, during or after or before, it doesn't matter, are the ones the ones that feel, the ones that can imagine the pain of others. The ones who see their mothers, their children, their sisters, in the face of the supposed enemy; the ones who feel their own pain in the way that men aren't supposed to feel, the ones that cannot suppress how they feel even after the order from their officers, *Suck it up, don't act like a cunt, where's your balls, man,* even after the slapping and corporeal punishments from their officers. The stubborn ones, who would not submit to the way a soldier is supposed to be, these are the ones that buckle under the weight of the nation, that lose their minds. The ones who aren't left to die on their own, losing their minds as much as their ways in the jungle, the handful few are scurried away like unwanted storage into the boat, away from the healthy ones, the mental illness known to be demoralizing and, sometimes, contagious, then into the selected mental hospitals on the Mainland when the ships were still running between here and there. They will stay in their rooms, some even sixty years after the war, treated as mental patients, because they are, because they are no longer fit to live in the society that cannot accept men collapsing under the weight of the war.

End of May, 1944: The Reduction of the Primary Platoon

Men have been left in the field hospital; one or two could not keep up with the march, telling Shirō, *I'll rest a bit, I'll catch up,* and that was the last he saw of them; one or two had fallen into the river, the last he saw was their arms thrust out of the fast current, begging to be helped. Now, only half a dozen of the original men were left, and if they were not weakened from malaria, from malnutrition, then from jungle fever, from their feet rotting during the march from the boots that did not fit them right, or the fractured back or leg bones from the weight they had to carry, the bones not withstanding all the weight hoisted up onto their backs. And new men, men he has never seen, put into the platoon because they were found wandering on a nearby path, lost from their original unit, hungry and delirious from their journey here. A rumor: there is going to be a major offense soon. A rumor: the government has abandoned them already. A rumor: there is no more food left at the supply base. A rumor: there will be a ship coming to get them to go home, released from their conscription with the words, *job well done*. Truth: there is no more food left, and even the Regimental Commander's horse was shot. Truth: the planes that had evacuated to the north, to the west, have all been shot down, even the pilots are earthbound, unable to fly, because there are no more planes left in New Guinea. A rumor. A truth that no one has access to. Always a rumor.

June 12, 1944: The Command to Mobilize to Wewak then to Aitape

The first command of offensive attack: Aitape. To wrestle it from the Americans and Australians. To trek again, back to Wewak they had left in December of last year. This is their first battle on this island. This is the first offensive attack and no one is ready. They are told not to take the path along the sea because the enemy has already taken it; they are told to go into the jungle then onto the mountain; they are told only to move during the twilight then at night when there are no enemy planes; they are told to move quietly.

June 25, 1944: Shirō's 28th Birthday

Shirō just turned twenty-eight today. He did not remember his birthday. The measurement of days has been uncertain; ever since the retreat, no one has been sure about what day of the week, what day it was. He is walking toward the Druinimor, or Bandō river, because the order came five days ago, through word of mouth, that all men of 237th Regimental Unit must gather there for the attack. The supply station had given each man a bag of rice, the first in many months. And he knew that this battle is meant to be the last battle. And this bag of rice, a last feast. He does not know that he will die in less than two weeks, three days before the major offensive, by the river named after the one behind his house back home.

End of June 1944: Shirō

How much he has forgotten. This all feels like a dream, this life he is living right now. The life before the New Guinea jungle, the life before on the farm, when he had just met Masa. How he had loved her from that first moment, and how he had felt, his hands calloused from holding hoes and things he loved holding: rackets, bicycle handles, bamboo swords. And there she was, her skin so pale it seemed as if the sun had avoided her since the moment she was born. It was the kind of paleness he had only seen in the paintings of beauties by the masters, her skin the color of the rice grain. So transparent, so frail. And she was educated, more educated than he was, having gone to a women's college. She had never known farming, and his parents had worried about that, but he knew she was the one. But was that all a dream? At the most startling moments, his two lives — the one here and the one he was forced to leave behind — collide, and he becomes disoriented, unable to distinguish which is his reality, which isn't. He is, even now, the same man. But not entirely. This is all he thinks as he sits in the jungle, his men taking a small rest, all holding each other's shirts as if they are children afraid of letting go, afraid of suddenly finding themselves swallowed up in a darkness so black and malicious that it is what Shirō imagines death would be like. The darkness of the jungle is all around him. So are the eyes he cannot see, but still knows are watching, sometimes benevolent and sometimes cruel. His men, what little of them are left, do not move; their sleeps are that deep, and perhaps that deadly. Who will wake up? Who will not? He closes his eyes. He will need all the rest he can render out of the days that are left. They must hurry. The messenger came so late, exhausted, collapsing as he handed them the crumpled message. They must reach the Bandō river. But right now, his men need all the rest they can muster in this dark jungle that does not yield rest or safety.

July 5, 1944: Shirō Preparing for the Battle of Aitape

It is dark. In the jungle of New Guinea, the sun never penetrates the ground, and because of this, trees grow fierce, gluttonous for the light, pushing each other to grab a morsel of it. There are eyes all around, but he cannot see who or what is looking at him. Birds that take on a vision of paradise when they open their wings — they are the colors of paradise, but when they cry, they cry the agony of an infant dying with fever, the gurgling that announces the last breath to come. After the long march through the jungle, he has lost many men already, and each loss weighs on him as if he were losing his own life. It is not a release he is feeling as he sheds one kilogram. No, it is taking him nearer to death, he knows that much. *Will they recognize me now?* he wonders in the moments when he is lucid enough to remember that this is not the reality of his life, when he remembers that he has a beautiful wife, three children, his parents, all waiting for him at the farm by the river, the village where people boated from one neighbor to another. His home. And in the moments like this, when he feels himself back in his emaciated body, he wonders, *will I come out of this alive?* He pushes these moments of doubts — these moments of lucidity — away. He must remain a soldier, a sergeant, a *role*. His job is to keep his men alive; his job is to do his duty. He must not become a man, for a man is one who has past histories. This doubt of mortality is what, at the end, kills a soldier in the battle. He makes his way through the thick, unyielding jungle with a handful of men who are still alive after eighteen months. And these men can barely walk: malaria — and various injuries that were never treated because there was nowhere to treat them — ravages their bodies. He knows, because when the order came to cross the river, to gather at the river for the attack, he understood that it was a foolish order, a desperate order, too fitting for the code name of his infantry regiment, *Kawa 3564* — if read out loud, *Mi-go-ro-shi* — or, *Left to Die*. He's never been lucky with the regiments he belonged to; the one in China was *Shi-ni-go-ro: Time to Die.*

July 8, 1944: Shirō's Death

It should have been an easy mission, a reconnaissance mission like any other he's been on: to find out the enemy position, and also to find out a strategic position for his unit. A small reconnaissance group chosen out each unit — five men in total. He is exhausted from walking non-stop for the past two weeks; he knows he is not ready, he has eaten the last of the rice he was handed two weeks ago, but this is a mission. He has been commanded to do so, and because he is an officer in the Imperial Army, because he is a sergeant, it is his duty. He moves stealthily through the dark jungle with the others, making sure not to make too much sound, as he has been trained to do. The river is in front of them; he sees lights, he hears laughter, he hears *English,* and they look at each other. Even in the darkness, he sees alertness in their eyes. And suddenly, the shrill, the familiar shrill, *Get down, get down!* He curls into the defensive position, but it is too late, something explodes by his ear and he shouts *I don't want to die!* But his body betrays him, it is no longer his will, no longer the spirit, that keeps him alive. At the very last moment of his life, the body accepts its own end, but not his soul. *I don't want to die,* he shouts. But the words do not come out, his mouth already dead, and his soul is about to fly away, but he's pulling it down, he's pushing himself down, back to the body that has taken him so far away from home, the body that has been faithful, like a dog, carrying him to this southern island, the body that should have carried him home to where Masa is, where his home is, *Not now,* Shirō shouts, *I have to go*

IMAGINARY DEATH

home, I have to go home. But it is already too late. And things slow down, just like they say, a second stretching into his entire lifetime. And suddenly, he feels himself calming. How strange, he thinks, how strange that the jungle, the very thing he feared and hated, is now offering solace. Why did he not see that the green of the jungle was the same color as the rice field in late May, or that the rain that came was as stark and as beautiful? As bountiful. He tries to find the moon in the sky, but there is no sky — only trees that offer no light or glimpse of the sky. His wife must be seeing the moon from the farmhouse. How he wishes he could tell her that now, as a reminder — that the Earth is small, and that they share so much of the same, though divided by two different realities — one of the war, another of the home front. How many times he has written to her, *though we are separated, our hearts are one.* Did he ever feel like that with anyone else, with the Emperor? No, he knows that she is the most important person in his life, that she is the one he has been fighting for all along: her beautiful skin, her quiet voice, her startling intelligence. He closes his eyes. He sees the rice field, golden, right before the harvest. The image of Seijurō flying a kite that new year so long ago. The way the wind blows over it, and the way the plants make such majestic music. Yoshiko and Matsuko sleeping, entwined like kittens. All these images, almost like the symphony. He inhales deeply and smells the jungle; but he no longer smells the thick vegetation, moist earth, decaying mud, but smells, as if he is there, rice paddies and the river behind his house that always smelled of the sea. Once the soul leaves the body, he will not fly to the monstrous shrine where all the war dead went and continue to go, but to that farm by the river, where white shy herons glide in the quiet sky, the land so green that even the sky looks green. Green. He can go home now. The war is finally over. He has been blown to pieces — he and another man. There is no body part to gather and cremate, there are no items for the living to pick up, to carry home, and only one person survives to tell this story nearly eighty years later.

履　　歴　　書					
氏　　　名		紫村　種郎		生年月日	大正5年6月25日
退職当時の		本　籍　地		茨城県行方郡潮来町大字大洲50番地	
		官　　職		陸軍曹長	
叙位・叙勲		勲七等白色桐葉章（昭和15年4月29日）			
年	月	日	任官・進級・昇給	記	事
昭和11	12	1	歩兵二等兵	現役兵として近衛歩兵第1連隊第3中隊に入営	
	12	2	28		第1期卒業
	3	8	歩兵一等兵の階級を与う	歩兵科幹部候補生に採用	
	6	10	歩兵上等兵の階級に進む	乙種幹部候補生を命ず	
自 12 至 12	10 11	12 11		歩兵第1連隊にありて支那事変勤務に従事	
	11	30	歩兵伍長	予備役編入　引続き近衛歩兵第1連隊に臨時召集	
自 13 至 13	1 6	1 16		歩兵第1連隊にありて支那事変勤務に従事	
	13	8	1	歩兵軍曹	近衛歩兵第1連隊留守隊に配属を命ず
	12	1		近衛歩兵第1連隊留守隊に配属を命ず	
		同日		第9連隊附を命ず	
自 13 至 13	12 12	1 31		近衛歩兵第1連隊留守隊にありて支那事変勤務に従事	
	14	3	25		歩兵第212連隊に転属
		同日		第9中隊に編入	
	5	7		北支派遣のため東京芝浦港出帆	
		13		青島港上陸	
		16		青島発	
		17		斉寧着　同日より同地附近の警備	
自 14 至 14	6 6	10 23		き号作戦による魯南討伐に参加	
	6	15		朝陽洞附近の戦闘に参加	
		21		金家荘南方高地附近の戦闘に参加	
自 14 至 14	6 7	27 5		魯西地区討伐戦に参加	

Fig. 5.5. Shirō's army resume, page 1.

年	月	日	任官・進級・昇給	記　　　　　　　事
自 14 至 14	7 7	5 29		定陶県城内にありて同地附近の警備
14	7	29		湾澤県城内に移駐　同日より同地附近の警備
自 14 至 14	8 8	23 26		大黄荘附近の討伐に参加
自 14 至 14	8 9	27 1		候垓孔庄附近の戦斗に参加
15	9	15	軍曹	勅令第580号により
自 15 至 15	9 9	17 21		梁山附近の戦斗に参加
自 15 至 15	9 11	24 26		コ号作戦に参加
	10	3		平橋鎮附近の戦斗に参加
		11		東溝鎮附近の戦斗に参加
自 15 至 15	10 10	15 16		興安墩附近の戦斗に参加
	10	16		草堰口附近の戦斗に参加
		17		上岡鎮附近の戦斗に参加
		19		上岡鎮附近の戦斗に参加
		20		七里港附近の戦斗に参加
	11	16		湾澤着　同日より同地附近の警備
自 16 至 16	3 3	15 19		王坰堆附近の戦斗に参加
16	6	11		湾澤発
		同日		斉寧着
				斉寧発
		16		青島港出帆
		20		宇品港上陸
		24		歩兵第57連隊留守隊着
		25		召集解除
		29		
18	1	5		臨時召集により歩兵第66連隊補充隊第5中隊に応召
		11		歩兵第237連隊整備人員として屯営発
		14		門司港出帆
		15		釜山港上陸

Fig. 5.6. Shirō's army resume, page 2.

年	月	日	任官・進級・昇給	記　　　　　　　　事
18	1	18		鮮満国境通過
		21		河北省衡水着　第5中隊附
自 18 至 18	1 2	21 22		歩兵第237連隊にありて支那事変勤務及び 大東亜戦役支那方面勤務に従事
	1	28		南方転進のため衡水発
		30		揚子江通過
		31		上海着
	2	23		上海港出帆
		同日		北支方面軍司令官の隷下を脱し第8軍司令 官の隷下に入る
	3	21		南洋群島パラオ島上陸
	4	26		パラオ島出帆
	5	1		ニューギニア・ウエワク上陸
		同日		第18軍の隷下に入る
19	7	8	曹長	ニューギニア・坂本川において戦死

上記に相違ないことを証明する。

平成15年11月4日

茨城県知事　橋本

Fig. 5.7. Shirō's army resume, page 3.

IMAGINARY DEATH

Fig. 5.8. Seijurō and Masa.

Fig. 6.1. Kōkichi in the 1960s.

Epilogue

July 13, 1944: One Life Is Gone, and Now, Another Takes Over

Shirō is gone, and where he left off, Kōkichi takes over because this is how a family keeps going — from first son to third son because the second is already dead. Saburō, the patriarch, is still alive, but the heir apparent was blown to pieces in New Guinea; but no one knows it yet. Susumu has been dead nearly a year now, and he's been properly mourned. Kōkichi is in the Vessel Engineer Corps, landing in Phuket in Thailand now. He does not know this yet: the land, the family, his oldest brother's wife, their children, the interrupted lives. Saipan and Tinian will fall soon enough, allowing Americans to build air strips, where fleets after fleets of B-29s will take off to firebomb cities in Japan for the next twelve months. One out of every four Okinawans will die in one of the fiercest land battles including civilians. And in August 1945, America will drop two bombs, one in Hiroshima and another in Nagasaki, bombs like no other that will cast a shadow over the next eighty years. All of this will be his burden to carry for the rest of his life. He does not know this, not yet.

September 2, 1945: Kōkichi in Malay

He lines up, him and the rest of the men. They receive the news that the war is over; that Japan has surrendered. Some cry in relief, some cry in regret, but underneath it all, the sense of emptiness: what have they been fighting for, all this time? That is the question no one can answer, and they will be asking that question for the rest of their lives, despite the willing answer, the formulaic answer already at the tip of the tongue: *We have fought for peace.* The flag, the symbol of unity, symbol of the warrior spirit, lies atop the damp wood. Kōkichi salutes sharply; men under his command salute, though there is no more war, and men are allowed to do what they want. But habits are hard to shake off, and their bodies have been molded into these habits for such a long time. A man pours gasoline taken from their ship. He strikes a match. The flag goes up in flames, and they watch, disbelieving, but with relief, that the flag will not fall into enemy possession. Then they surrender themselves to the enemy. They walk up, one by one, amidst the jeering of the enemy, *Jap, monkey,* through the spit, and the catcalls, and the humiliation, to the pile where the rifles lie, to where the swords lie, and Kōkichi unbuckles his sword, the one he asked Shirō to procure for him, the one he was so proud to carry as a symbol of his officerhood, the representation of his years in the war, and lets it go. He is now a nameless Jap, a number, a POW.

September 1945: Masa

She reads the newspaper every day, for the news of men coming home now that the war is over, but all she reads is news of changes: war criminals arrested, but not the Emperor; prices changing overnight; cities in ruins, still, and not enough medicine, not enough water, not enough for people living there; all station names must be in English; Americans are our friends; women are told never to smile at Americans or else, or else; Hiroshima and Nagasaki are safe to travel to; the weather column is back. The State is changing overnight, and people talk of democracy, *How good it is we have lost the war;* people are changing overnight, and she is startled to find that so many who had only spoken the rhetoric of the nation, *Die honorably, kill as many enemy as you can, American and British devils,* the words that had only a month ago littered the newspapers that encouraged people to spear, to kill, to butcher the enemy when they landed, were speaking new words: *democracy, freedom, liberation.* How the world has changed. She shakes her head, but she is looking for that one line, somewhere, anywhere in the newspaper: *18th Army from the South is being repatriated.*

January 9, 1946: The First Repatriation from New Guinea

Saburō has been obsessively listening to the radio ever since the war ended in August. He's heard so many rumors, not on the radio but by word of mouth, rumors that started out with someone's question, *I wonder when our men will come home?* An innocent question that turned, at the moment of exasperation, into a statement, a fact, truth, as it went from a mouth to an ear to another mouth. Rumors that dictated his life now. His sons must be coming home, they must, he thinks, his eldest son in New Guinea and his third one somewhere. He has already given up on his second son, the one who died in the sea, because the government said so, because they got the notice back on July 4, 1943, the one he had the Buddhist priest rename *Truly Loyal Brilliant Sunlight Susumu,* because he was loyal, and because he was brilliant like the sun. And today, the news on the radio that there will be five ships that will bring back former soldiers from New Guinea, all arriving later this month. His heart jumps in a way it has not in a while; he runs around the house, then out to the field. *Shirō is coming home! Shirō is coming home!*

January 18, 1946: *Kōei Maru* Comes into Uraga Harbor, a Former Great Warship Now Carrying Former Soldiers

Not to Ōtake Harbor, that is too far, but to Uraga, that is where Saburō can go. He takes with him a relative, maybe a cousin, he has ridden so many hours, two days, changing from one train that halted mid-track because of black market raids, another train where men and women hung from windows, people crammed into slow trains, through the devastated ruin that used to be Tokyo, then to another train through an equally devastated landscape, for two days, for three days, just so he can be there for the ship to unload its contents. First, men on stretchers come down, carried by men from the dock, cared for by nurses. A stark contrast between these men's uniforms, tattered, the original color no longer detectable, and the nurses' crisp white uniforms. These men can barely walk; they can barely stay alive. Saburō does not know that the nine-day trip across the ocean left a trail of men dying on the ship, their bodies thrown into the ocean before they could reach the home that they longed for. Saburō scans from one face to another, but he cannot see Shirō, he cannot see the familiar face of his son, but was it because all these men do not look like men anymore but like the mummies that he's read about? Is it their eyes that looked more dead than alive? And after them follow men who can walk on their own, but barely, men who come down, limping, being helped by their buddies, who hobble down the ramp, their legs so skinny their knees are bigger than their thighs. *Has anyone seen Sergeant Shirō Shimura? Has anyone seen...* but his words are shadowed by others' voices, yelling, *Have you seen ——? Have you seen my son?* And he is told by one of the ghost-like soldiers coming down the ramp, *I think that unit is coming on the last one....* Something in Saburō breaks, maybe it's akin to his back dislocating, this sense of losing strength in his legs, as if his body is no longer his own, but this will not be the first one or the last one.

January 31, 1946: The Last Ship Arrives from New Guinea

Saburō waits at home this time. The last trip, the one he made two weeks ago, was too much; the unreliable and long train ride, train tickets that tripled in price, the stress. He is an old man, and who knows, Shirō might have arrived on a different boat that made port at Ōtake. Who can trust what a former soldier says; who can trust rumors anymore? He waits. He waits. He does not sleep that night; any minute, at any hour, a telegram can arrive, a telegram from Shirō, a telegram from the hospital because he is sick. He waits. He does not know that out of 160,000 men sent to New Guinea, 150,000 died before the war ended. Out of the 4,194 men in his regiment, only 44 made it alive to the homeland.

1946: After the War, the Quiet Homecoming

The war ended a year ago. The soldiers began to come home soon enough from the *outerland*. If they had not been captured, they came back on the Red Cross boats. If they had been captured, they came from the prisoner-of-war camps on the southern islands, from China, from Southeast Asia, even though there was no more war, so technically they were no longer prisoners-of-war. They crossed the ocean not as citizens, but as repatriated soldiers of a defeated nation; they landed on the *innerland* to find the landscape flattened from the air raids, and people walking around with their faces toward the present, not the past. The present: the immediate lives they had to live. Instead of finding the nation waiting for them, soldiers came home to men and women looking at them with indifference; they did not know that indifference was a kind gesture until they came back. Now, the moment they stepped down from the train, men and women seemed to hiss; the whole nation was hissing at them: *You lost the war. It's not us, but you, who were so weak. You probably didn't fight until the death. Why are you alive when our sons, our husbands are dead? You are cowards, you probably left our fathers to die and surrendered.* The mass that had sent them off as heroes, as future *hashira* — pillars — to serve the Emperor in life and in death, now, those very same people spat on them, ignored them, blamed them for the shame so heavy it even brought down the Emperor from his rightful throne in heaven. Soldiers returned to the fifteen years of grief the nation could not express, the very grief that was silenced during the war in the name of the Empire. Now, as if waking up from a dream, the mass shattered. Each fragment became an individual. And as an individual, they pointed fingers at these men who failed miserably. Soldiers came home to an unfamiliar land.

June 21, 1946: Kōkichi Leaving a Malay Prison

He did not know until now that men would do anything to stay alive, even betray each other. One interrogation after another of what they did with prisoners-of-war during the war. One man under him says that he, Kōkichi, commanded the beatings, and that he had no other choice but to beat the prisoners or Lieutenant Kōkichi Shimura would have killed him. One man says that another officer commanded the killings, and that Kōkichi was an accomplice, but Kōkichi suspected that this man said it, parroting the answers *they* wanted to hear, so that he could get a pack of cigarettes from the former enemy. Kōkichi says that he was an officer in the Vessel Engineering 15th Regiment, all he did during the war was on the boats, responsible for shoring up supplies to men needed. *But we have it on the record that your regiment was responsible for guarding POWs....* And he tries as best as he can in his English, No, our duty was to oversee... he does not trust the interpreter who speaks strange and halting Japanese. Interrogation after interrogation. Then work around the camp until the body gives up, and it gives up easily with the food handed out in the morning, at the end of the day. Not enough food to sustain the labor. People betraying each other; men dragged to a make-shift court room, judged, sentenced, based on a confession of another. Kōkichi does not say anything.

EPILOGUE

We know you were... until one day, a document is read to him: *He came into my cell and gave me alcohol, saying that I must be cold; he saved my life.* He is puzzled. He does not remember this. *Who is this man?* And a name is read aloud, but there is nothing, no memory of this incident. Kōkichi does not remember, then yes, somewhere, he remembers that night: he was drunk that night, he was bored, he wanted to speak English, he wanted something different, so he went to a cage — the one where they kept the American or Australian or Dutch, he did not remember, the one others were having trouble with because he was defiant, that POW, stubborn and proud — and he unchained the man. He gave him some alcohol. He said something in English, maybe, *Good? You like?* Because English had been banned for so long, his tongue did not move as quickly as it should, these words hiding shyly in the recess of his memory. But soon, as he and the POW shared the bottle, alcohol loosened the tongue, and he was speaking his best English. He does not remember what this POW and he talked about, maybe about where he was from, maybe he even told this foreigner his name, maybe how he has two brothers, maybe... it is unclear. It was not kindness, but boredom, maybe a little curiosity as well, but a simple gesture that came from his everyday self, the self that had been fighting against the system, that had defied these moments, and something so inconsequential in his mind is a defining moment for that man, a man he doesn't even remember the name or the face of. He is let go; other prisoners accuse him of naming names, but he knows better. He did one simple thing, so little he did not remember it, something ordinary he would have done if it were another time and another place, but rendered extraordinary during the last fifteen years of this war.

July 1, 1946: The Return

When Kōkichi left in January 1941, as if changing places with Shirō who came home on June 1940, he left amidst the fanfare, a proud third son, a man off to basic training. And when he got into the officer's school, he left amidst another fanfare, a proud son, with a new sword glistening beside him. He spent the last two years on a boat, from Korea to Manila, Cebu Island, Vietnam, Burma, Thailand, Singapore, never staying in one place long enough, but transporting food and supplies and sometimes enemy prisoners. His regiment did not know about the surrender in August 1945; instead, he was promoted to First Lieutenant on August 20, and 13 days later, on September 2, his regiment put down their arms. And here he is now. They have taken everything from him, first at the surrender, and then at the prisoner-of-war camp and the war trials, and then again on the boat back home after his release. The only thing that they allowed him to keep was the malaria that he had picked up somewhere along the way and gave him a burning fever and chill every other week. He would spend the next month on the boat, praying to stay alive, just for that one step toward home, for malaria to go away, while all around him, men kept dying and were released into the ocean. *Almost home,* he keeps telling himself. *I have to stay alive, at least until I get home.* He talks of home, of what he would do when he gets home, and men all around him do the same. It is a prayer, this talk of home, as if talking about it will keep them alive, will allow them to see it. In front of them, the Ōtake Harbor in Hiroshima. Hiroshima completely flattened out, still, even after a year. Charred, gnarled, inhabited by people as gnarled as the city itself. He cries; men all cry. They are home, finally. They're home, but why do they feel guilty? What is this land they have landed on? Confusion. Joy. Sadness. Guilt. All of that rolled up into heartbreak. He loses words, words stuck in the roof of the mouth as he stumbles down the ramp onto the harbor, as the repatriation officers spray him and the surviving men with DDT from head to toe, again and again. He is given a free ticket to go home and some money—new bills that look like toy money—to get home. They sign themselves back to Japan; he sends a telegraph home, telling them he will be back soon. No one looks at the recently repatriated soldier; one woman even shoots him a look of hatred, as if to say, *What are you doing here?* Here he is, 25 years old, having spent the last five years at war. He does not know, yet, that he is the only one of his brothers to survive the war, and that he is now the eldest son, the man who will carry the burden of the family.

Mid-July 1946: Yoshiko's First Complete Memory

This is her first complete memory. Yoshiko plays with her younger sister, Matusko, in the river. It is a month away from the harvest time, the landscape green, green with rice, green so thick the land itself is green. White herons glide down, then land, and fly away. The house itself is surrounded by a dark green, late-summer rice field — the river she stands in is shallow, gentle enough for children to swim in, because it is late summer and there has not been rain for a month. But the house is surrounded by rice fields on three sides, with the river running behind. She looks up, and in the far distance, she sees a man coming toward the house. No, it is two men. A man walking slowly, hesitatingly next to another man pulling a bicycle. They stop, and the bicycle man leans close to the walking man as if he is asking a question. Their faces are unrecognizable from the way the man's back is bent forward from the rucksack, or is it from exhaustion? And a tattered hat that shades his face. From afar, where she is, Yoshiko knows that the man is a soldier, a figure clad in beige like any other soldier who had been released from the Surrender, just like all men passing through the village, all with the same look on their faces: bewildered, released, dejected, eyes barely visible in unnaturally haggard and tanned, loose skin, their eyes shaded as if avoiding the sun, as if the sun would light up on the secrets and shame in their hearts. The sun beat heavily on her neck, but Yoshiko only hears a keening scream, and as if led by the sound, a heron takes flight. Then hurried steps, straw sandals slapping against the ground, feet kicking the earth. Yoshiko and Matsuko look at each other. They hurry up the bank to get a better view, keep looking at the two men approaching her world. She does not know that her world is changing, and it changes.

August 1946: Homecoming in Bed

He shakes under the thick blanket, the malaria fever ravishing him. No matter how hot it is out, no matter how many blankets are piled atop him, he is still at war, wandering on the dark sea as he had done so many nights, and he feels that fear he had carried all those times: torpedoes. Dying in the sea, when no one knows his vessel has been sunk, alone. Or even worse, surviving in the sea, afloat, knowing that any minute now, his strength will give out, or that a shark will tear his limbs one at a time. And suddenly, he is back in Korea, singing with the prostitutes the only Korean song he knows — it may have been fake for all he knows,

Arirang, arirang, arariyo, ariang gokyelo
namwam kanda, nalul burigo gashinun nimum,
shimulido mokgasaw balbyuhng nanda

Ariran, ariran, araiyo, ariyan,
the one who crosses over the ridge, the one who is leaving me
gets sick within ten li, and he is sick.

He is sick. He is home, he is not home. The fever rushes through him, then the cold, and he is in southeast Asia, in Thailand, and he is on Attu Island where he has never been, he stands on the vessel at night, he sits in his cell, he is at the officer's school writing to his brother, *Please send me…* and he is standing in front of his students, teaching… and he sings *arirang arirang,* to keep still so he does not shift in and out of time, so he can stay, and *arirang…* and suddenly, in a fraction of a second, he is drawn back to his own body, he is here, at home by the river, and the door slides open. He opens his eyes and he hears the world of his family.

EPILOGUE

September 16, 1946: The Official Notice Arrives

A man in a black robe walks toward the house, once in a while stopping to mop off his sweat. He is an old man, and from the way he is walking, in small steps because of the robe, but slowly, deliberately, it is plain that he is the village monk. In the rice field, after the harvest, the earth hardens from the stopped water. No one is home except for Kōkichi lying on the rattan chair, minding the little ones; he watches the monk, then he watches his nieces in the yard, laughing in their imaginary world, and he envies them because he cannot escape the past. Even when the monk is by the gate, Kōkichi does not get up as befits the monk; in truth, he still does not have the energy to stand. He cannot tell whether it is the malaria or just the shock of being home. When he heard about Susumu's death between his bouts of fever, he did not feel anything, like he didn't feel anything when his little vessel tried to drop off much needed food and ammunition on a small island in the middle of the night, and a man waded through the shallow toward the vessel, whispering, *Take me with you, I can't stay here anymore, I'll go crazy, take me,* but Kōkichi followed his command: *Do not bring anyone aboard. Drop off the supply and leave as quickly as you can.* That's what he did, coldly, but in his mind, he kept apologizing, *I'm so sorry, I'm so sorry.* The monk is standing in front of him. He does not look at Kōkichi; he just looks at the ground. The little ones have stopped what they are doing and are smiling at the monk, expectantly, but the monk ignores them. The monk is sweating; his bald head glistening from the sun reflected on the sweat. *I'm so sorry, but your oldest brother has arrived at my temple....* Kōkichi's heart leaps, but the monk, in a quieter voice, says, *his remains....* Kōkichi does not hear anything else. He sees the monk's wrinkled lips moving, but no word comes out of them; he feels the tingle of his toes, the roaring in his head, as he would at the onset of every bout of malaria; he is waiting for that rush of heat to lock his joints, to bind him, but it does not come. He keeps looking at the monk's mouth but he does not hear anything.

Grief Contained

Grief cannot be compacted into words, it resists being rendered into language. Here is a man who is dead. Nothing is contained in the urn that carries Shirō's name. There is no informal parting demanded between the dying and the living; there is only formal departure, named by the state. Nothing made it out of that island, only 44 men out of 4,400 in his regiment. Not the notebook of the names of the dead, because the survivors had to relinquish everything personal, even the photographs that the dead may have given them, all the letters, all that made them individual, when they surrendered and entered the camp. Not the memory of how Shirō died, because most of the men who might have known his last moments were dead, and along with their deaths, the story of how Shirō spent his last moment alive. The way the news of all war deads' last moments was transmitted: repatriated soldiers coming off that boat were required to tell the men from the Repatriation Department what happened to their platoon, unit, company, regiment. They were asked about anyone they might have known. They were asked about their brother-units, the ones that fought with them. And the men from the Department tallied up all the narratives, then made subtractions: since no one in that unit survived, everyone must be dead; someone told someone about Shirō's unit, his regiment, and with no trace of him after the Battle of Aitape, his death was declared: he must have died during the battle. One document says that he died two days before the offense; another says that he died on July 19th during the battle. But without proof, without someone who saw his last moments, without the body, how can grief start? How can anyone trust words, rumors, lies wrestled into truth, after all these years, especially the words of the State that had been lying for such a long time?

EPILOGUE

October 19, 1948: Marriage of Kōkichi and Masa

They have waited two years now, but the family must move forward. *I could have married anyone; there were so many single women, almost a truck-ful for each man,* Kōkichi would say later when he was drunk and his words slid easily out of his mouth. They waited two years, but it wasn't long enough; they've heard enough stories of men who were declared dead coming back two years later, three years later, only to find their wives married to someone else. Married to the surviving brothers. But two years. That is long enough. Kōkichi had tracked down Shirō's commanding officer — the only one to survive of that unit — and though his words were minced, he knew that Shirō was dead. Maybe a rumor, maybe someone else. But the government declared Shirō dead. That was enough. Yoshiko remembers all the neighbors coming to her house. The house that had been quiet for such a long time because there were no occasions to celebrate. For such a long time. She remembers looking in from the partially closed sliding door, listening to people laughing, eating, people being happy. Masa is five years older with three children by his brother. Kōkichi and Masa will have a child of their own next year, and they name her Sachiko, a child of happiness. As much as Yoshiko liked her uncle who turned into her stepfather, she never forgave him for saying, whenever he was drunk, *I got a hand-me-down wife. I always got my brother's hand-me-downs.* Even in her late 70s, when she becomes older than when both Kōkichi and Masa died, she didn't forgive.

1950: Kōkichi and Matsuko

Kōkichi hoists Matsuko onto the back of his bicycle and ties her tightly. He tells her gently, *Hold on, try to hold on, you don't want to fall, do you?* This is his good day; he is not shaking from the malaria fever that comes at unexpected moments, when he feels the wave of ice rushing from head to toe suddenly, then the blast of heat so hot he can't stand still, when he would have to bound himself under blankets, layers and layers of blankets, but still shaking from the cold, singing a Korean song he learned so long ago. This is a good day, but it isn't for Matsuko. She can barely sit up; he keeps feeling her sliding off, and he would stop the bicycle to readjust her every five minutes. They are on the way to get a shot of penicillin for her, twice a week, a shot that costs a thousand *yen,* and she needs two a week. This is when 30 kg of rice costs the same, but Kōkichi has been barred from any job because he has been a lieutenant in the army that has not existed for five years now, and he cannot get a pension because even though he gave five years of his life, the government he served under no longer exists. But he had promised his brother: he would look after the family if he survived the war. And he does.

October 17, 1950: Giving Up the Dead

It is easy to give up the dead, to release them from the property of the family to the State when the death is definite, a certain death that offers no conjuncture and no possibility. Eiko saw Susumu in a dream when his ship sank — why else would he, at the moment of his death, travel such a great distance over the sea to say goodbye? Susumu was dripping wet, fit for the drowned dead just like the condolence letter said: *Died in the battle of the Bismarck Sea,* and he said, without moving his lips, how he did not want to die, how he regretted that he had to die. So much more to do, he would have said, if there was more time and if the spirit could have all the time in the world to talk. It was a certain death. So Saburō submits Susumu's name to the Yasukuni Shrine, to enshrine him as a *pillar,* to release him to the State so that he will forever be remembered, because there is no one who would grieve for his death except for his parents and his siblings. After all, spirits of those whose lives end abruptly, without having lived a full life, haunt the living. They become restless; they turn childish, demanding attention. And that is not what the living want. That's what the Shrine is for, to quiet the restless spirits of the war dead. Turn them into gods instead of curses. And after his parents die — and parents must die, eventually — who will remember him? Who will remember a man — one of thousands and thousands — who died alone in the sea? So instead of the singular family, singular story of a man, Saburō gives Susumu up to the State, where no one will remember the details, no one would remember how much Susumu loved swimming, how he loved playing soccer, how he fought against the system, how he escaped from the farm to forge his own life, how he wanted so much out of life but instead got caught up with the fate of the nation. The State accepts, though they misspell his name and write down the wrong date of death, swallowing him into the masses of dead, his story turning into the narrative of the nation, *How many brilliant lives were cut short because of the war.*

Seven Years after the War

Kōkichi tills the land. He has been doing that for six years, waking up every morning to go out in the field, working until lunch time, then until it is dark. On his good days. But on bad days, he lays shuddering in his bedding, unable to think straight, unable to keep himself warm, even after all these years. He carries the southern island sickness after six years. He also carries the land on his shoulders and his oldest brother's family — three children and a wife — as well as the entire weight of the family line. Men with pasts like his — officers in the former Imperial Army — have been banned from working. Gorō, Masa's younger brother, the ace pilot turned experimental pilot, the very pilot that the Army refused to send as a Tokkō pilot, is now working in the untamed land right by the sea, making salt out of seawater. Kōkichi has been banned from going back to teaching, and he has become a farmer, an unexpected life before the war. But the war changed everything. His brothers are dead. His parents hardened into grief. He is there when Saburo dies of a stroke, and he is there, sleeping next to Eiko as she spirals into the past, waking up at odd hours when she calls to him, calling him *Susumu, Shirō,* most of the time not recognizing him, and, like a dutiful son, only a few weeks after Eiko died, he clutches his chest in his office at an elementary school where he is the principal. *I can't die yet,* he is said to have said. *I can't die yet.* And he dies of a heart attack, though Yoshiko says that he was just like his name, *The Son Who Serves,* accompanying his mother through life and into death. He had promised his oldest brother that he would come out alive, and he did; he promised that he would look after the family, and he does.

履		歴		書	
氏　　名	紫村　孝吉			生年月日	大正１０年５月１４日
退職当時の	本籍地	茨城県行方郡潮来町大字大洲５０番地			
	官　職	陸軍中尉			
叙位・叙勲	正八位（昭和18年12月28日）				

年	月	日	任官・進級・昇給	記	事
昭和17	1	10	二等兵	現役兵として歩兵第102連隊補充隊第8中隊に入営	
	4	23		歩兵第2連隊要員として屯営出発	
		24		宇品港出帆	
		26		釜山港上陸	
		28		安東通過	
	5	1		北安省嫩江県嫩江着	
		同日		第10中隊編入	
	7	10	一等兵		
	8	20		昭和17年度採用兵科幹部候補生を命ず	
	10	10	上等兵の階級に進む	甲種幹部候補生を命ず	
	11	1		仙台陸軍教導学校に分遣のため嫩江出発	
		4		安東通過	
		7		釜山出発	
		同日		下関上陸	
		10		仙台陸軍教導学校に入校	
	12	1	伍長の階級に進む		
18	2	1	軍曹の階級に進む		
	4	28		同校卒業	
		同日	曹長の階級に進む	見習士官を命ず	
		29		仙台出発	
	5	1		下関出発	
		同日		釜山上陸	

Fig. 6.2. Kōkichi's army resume, page 1.

IMAGINARY DEATH

年	月	日	任官・進級・昇給	記　　　　　事
昭和18	5	4		安東通過
		6		黒河省嫩江着
		同日		第10中隊に配属
	10	1		士官勤務を命ず
		9		第12中隊に配属
		13		神武屯に追及のため嫩江出発
		14		神武屯着
	11	30		現役満期除隊
	12	1	少尉	予備役編入引続き同隊に臨時召集により歩兵第2連隊応召
		同日		歩兵第2連隊付
		4		神武屯出発
		5		独立工兵第55大隊に転属
		同日		嫩江着
		11		黒河省嫩江出発
		15		鮮満国境通過
		18		釜山着
		同日		第4中隊付
		19		釜山出帆
		同日		門司寄港
		21		門司出帆
		24		高雄寄港
		25		船舶工兵第15連隊に転属
		26		高雄出発
		28		比島マニラ港着
19	1	2		マニラ港出帆
		4		セブ島セブ着
		5		セブ出帆
		7		ネグロス島ドマゲテ着
		8		軍令陸甲第112号により昭和18年12月25日付け船舶工兵第15連隊要員として同地において転属

Fig. 6.3. Kōkichi's army resume, page 2.

月	日	任官・進級・昇給	記　　　　　　　　事
1	8		第2中隊附き
2	23		馬来派遣のためドマゲテ出帆
3	1		マニラ港上陸
	14		マニラ港出発
	17		仏印西貢上陸
	26		西貢出帆
4	1		昭南上陸
5	20		ペナン移駐のため昭南出発
	29		ペナン島グルゴ上陸 同日より同島付近の警備
7	2		作命甲第83号に基づき「プライープーケット」間海上輸送のためプライ出帆
	同日		泰国境通過
	13		プーケット上陸
	28		帰隊のためプーケット出帆
8	5		泰国境通過
	6		ペナン島着 同日より同島付近の警備に従事
9	18		ペナン出発
11	6		泰国「カオハージ」着
	28		カオハージ発
12	1		泰国境通過
	2		緬甸国メルギ着
	3		プーケット付近局地海上輸送のためペナン出帆
	同日		泰国境通過
	9		泰国プーケット着 同日より同地付近の警備並びに海上輸送作業に従事
1	3		帰隊のためプーケット出帆
	同日		クラビー着　原隊復帰 同日より同地付近の警備
6	28		メルギ出発

Fig. 6.4. Kōkichi's army resume, page 3.

IMAGINARY DEATH

月	日	任官・進級・昇給	記　　　　　　　　　事
	10		カオハージ出発
8	1		泰国カンタン着
8	20	中尉	
	25		海上輸送第7大隊に転属
9	2		泰国において武装解除
6	21		マライ出発
7	1		大竹港上陸
	同日		復員

に相違ないことを証明する。

成19年2月9日

茨城県知事　橋　本

Fig. 6.5. Kōkichi's army resume, page 4.

Bibliography

Akahata Society Page Editorial Department 赤旗社会部編. *Shōgen Teikoku Guntai* 証言帝國軍隊 [Confessions: The Imperial Army]. Tokyo: Shin Nihon Shinsho, 1989.

Aldrich, Richard J. *The Faraway War: Personal Diaries of the Second World War in Asia and the Pacific*. London: Corgi, 2005.

Antoni, Klaus. "Momotarō (The Peach Boy) and the Spirit of Japan: Concerning the Function of a Fairy Tale in Japanese Nationalism of the Early Shōwa Age." *Asian Folklore Studies* 50, no. 1 (1991): 155–88. DOI: 10.2307/1178189.

Aonuma Yōichirō 青沼陽一郎. *Kikansezu: Zanryū Nihonhei Rokujūnenme no Shōgen* 帰還せず 残留日本兵60年目の証言 [Unrepatriated: Confessions of Japanese Soldiers after 60 Years]. Tokyo: Shinchosha, 2006.

Ara Kenichi 阿羅健一. *Nankinjiken Nihonjin 48nin no Shōgen* 「南京事件」日本人48人の証言 ["Nanking Incident": Testimonies of 48 Japanese Eyewitnesses]. Tokyo: Shougakukan Bunko, 2002.

Arai Shin'ichi, ed. 荒井信一編. *Nijū-Seiki no Sensō: Nicchu Sensō I to II* 二十世紀の戦争IとII 日中戦争 [War of the Twentieth Century: Sino-Japanese War Volumes I and II]. Tokyo: Kusanone Shuppan Kai, 2001.

Arai Tomiyo 荒井とみよ. Chūgoku Sensen wa Dō Egakaretaka: Jūgunki o Yomu 中国戦線はどう描かれたか 従軍記を読む [How the Chinese Theater Was Portrayed: Reading War Correspondence Literature]. Tokyo: Iwanami, 2007.

"Attsu-to Gyokusai no Higeki: Yamasaki Yasuyo to Higuchi Kiichiro"「アッツ島玉砕の悲劇〜山崎保代と樋口季一郎」[Tragedies of the Battle of Attu and *Gyokusai*: Yamasaki Yasuyo and Higuchi Kiichiro]. *Web Rekishi Kaido,* May 29, 2017. https://rekishikaido.php.co.jp/detail/3934.

Bandura, Albert. "Moral Disengagement in the Perpetration of Inhumanities." *Personality and Social Psychology Review* 3, no. 3 (1999): 193–209. DOI: 10.1207/s15327957pspr0303_3.

Baum, Steven. *The Psychology of Genocide: Perpetrators, Bystanders, and Rescuers.* Cambridge: Cambridge University Press, 2008.

Baumeister, Roy. *Evil: Inside Human Violence and Cruelty.* New York: Henry Holt, 1999.

Baumeister, Roy F., and W. Keith Campbell. "The Intrinsic Appeal of Evil: Sadism, Sensational Thrills, and Threatened Egotism." *Personality and Social Psychology Review* 3, no. 3 (1999): 210–21. DOI: 10.1207/s15327957pspr0303_4.

Benedict, Ruth. *The Chrysanthemum and the Sword: Patterns of Japanese Culture.* New York: Mariner, 2005.

Berkowitz, Leonard. "Evil Is More Than Banal: Situationism and the Concept of Evil." *Personality and Social Psychology Review* 3, no.3 (1999): 246–53. DOI: 10.1207/s15327957pspr0303_7.

Berreman, Joel V. "Assumptions About America in Japanese War Propaganda to the United States." *The American Journal of Sociology* 54, no. 2 (1948): 108–17. DOI: 10.1086/220289.

Bischof, Gunter. "Victims? Perpetrators? 'Punching Bags' of European Historical Memory? The Austrians and Their World War II Legacies." *German Studies Review* 27, no. 1 (2004): 17–32. DOI: 10.2307/1433546.

Bourke, Joanna. *An Intimate History of Killing: Face to Face Killing in 20th Century Warfare.* LaVergne: Basic Books, 1999.

Brown, Stewart. "Japan Stuns World, Withdraws from League." *UPI*, February 24, 1933. https://www.upi.com/Archives/1933/02/24/Japan-stuns-world-withdraws-from-league/2231840119817/.

"Burial in the Aleutians." *TIME Magazine* 41, no. 26 (1943): 56.

Buruma, Ian. *Wages of Guilt: Memories of War in Germany and Japan*. London: Virago, 1994.

Cameron, Craig M. "Race and Identity: The Culture of Combat in the Pacific War." *The International History Review* 27, no. 3 (2005): 550–66. DOI: 10.1080/07075332.2005.9641072.

Chara, Paul J., Jr., and Kathleen A. Chara. "Kamikaze Attack Survivors: How Accurate Are Their PTSD Reports?" *Psychological Reports* 105, no. 3 (2009): 1126–30. DOI: 10.2466/PR0.105.F.1126-1130.

———. "Posttraumatic Stress Disorder Among Survivors of a Kamikaze Attack." *Psychological Reports* 89 (2001): 577–82. DOI: 10.2466/PR0.89.7.577-582.

———. "PTSD and Coping Resources Among Survivors of a Kamikaze Attack: The Role of Character Strength." *Psychological Reports* 95 (2004): 1163–71. DOI: 10.2466/pro.95.3f.1163-1171.

———. "Survivors of a Kamikaze Attack: PTSD and Perceived Adjustment to Civilian Life." *Psychological Reports* 99 (2006): 971–80. DOI: 10.2466/PR0.99.3.971-980.

Chūgoku Kikansha Renrakukai 中国帰還者連絡会訳編, trans. and ed. *Kakusei: Nihon Senpan Kaizō no Kiroku* 覚醒: 日本戦犯改造の記録 [Awakening: Record of Changes of Japanese War Criminals]. Tokyo: Shinpu, 1995.

———. *Shinryaku, Gyakusatsu o Wasurenai Tennō no Guntai Nihonjin senpan no shuki Dai ni shū* 侵略、虐殺を忘れない 天皇の軍隊＜日本人戦犯の手記＞第二集 [We Will Not Forget the Aggression and Massacre: Emperor's Army—Confessions of Japanese War Criminals, Vol. II]. Tokyo: Nihon Kikanshi Shuppan Center, 1989.

———. *Tennō no Guntai: Chūgoku Shinryaku Nihonjin Senpan no Shoki kara* 天皇の軍隊＜中国侵略＞: 日本人戦犯の書記から [Emperor's Army—Invasion of China: From the Records of Japanese War Criminals]. Tokyo: Nihon Kikanshi Shuppan Center, 1988.

"December 8, 1941 Ambassadors Nomura and Kurusu Special Envoy Hold a Meeting with US Secretary of State Hull to Deliver Their Ultimatum." *The US-Japan Talks as Seen in Official Documents.* https://www.jacar.go.jp/english/nichibei/popup/pop_29.html.

Dower, John W. *War without Mercy: Race & Power in the Pacific War.* New York: Pantheon, 1986.

Earhart, David C. "All Ready to Die: Kamikazefication and Japan's Wartime Ideology." *Critical Asian Studies* 37, no. 4 (2005): 569–96. DOI: 10.1080/14672710500348463.

Eckstein, Gustav. "The Japanese Mind Is a Dark Corner." *Harper's Magazine* 185, no. 1110 (1942): 660–68.

Evans, Michael, and Alan Ryan, eds. *The Human Face of Warfare: Killing, Fear, and Chaos in Battle.* Sydney: Allen and Unwin, 2000.

Ferguson, Niall. "Prisoner Taking and Prisoner Killing in the Age of Total War: Towards a Political Economy of Military Defeat." *War in History* 11, no. 2 (2004): 148–92. DOI: 10.1191/0968344504wh291oa.

Field, Norma. "War and Apology: Japan, Asia, the Fiftieth, and After." *Positions* 5, no. 1 (1997): 1–49. DOI: 10.1215/10679847-5-1-1.

Fletcher, Mark. "Intellectuals and Fascism in Early Showa Japan." *The Journal of Asian Studies* 39, no. 1 (1979): 39–63. DOI: 10.2307/2053503.

Ford, Douglas. "British Intelligence on Japanese Army Morale During the Pacific War: Logical Analysis or Racial Stereotyping?" *The Journal of Military History* 69 (2005): 439–74. DOI: 10.1353/jmh.2005.0089.

Francis, Timothy Lang. "'To Dispose of the Prisoners': The Japanese Executions of American Aircrew at Fukuoka, Japan, during 1945." *The Pacific Historical Review* 66, no. 1 (1997): 469–501. DOI: 10.2307/3642234.

Fujii Tadatoshi 藤井忠俊. *Heitaichi no Sensō: Tegami, Nikki, Taikenki o Yomitoku* 兵たちの戦争 手紙・日記・体験記を

読み解く [Soldiers' War: Reading and Analyzing the Letters, Journals, and Autobiographies]. Tokyo: Asahi Shimbunsha, 2000.

———. *Kokubō Fujinkai Hinomaru to Kappōgi* 国防婦人会 日の丸とカッポウ着 [Committee of Women's Defense: The Rising Sun and the Apron]. Tokyo: Iwanami Shinsho, 1985.

Fujita Masao 藤田昌雄. *Gekisenjō Kōgun Urabanashi* 激戦場 皇軍うらばなし [Fierce Battlefields: Untold Stories of Imperial Soldiers]. Tokyo: Kōjinsha, 2006.

Fujiwara Akira 藤原彰. *Nanking no Nihongun: Nanking Daigyakusatsu to sono Haikei* 南京の日本軍 南京大虐殺とその背景 [Japanese Army in Nanking: Nanking Massacre and Its Background]. Tokyo: Ootsuki, 1997.

———. *Okinawasen: Kokudo ga Senjō ni Nattatoki* 沖縄戦：国土が戦場になったとき [Battle of Okinawa: When the Mainland Becomes Battlefield]. Tokyo: Aoki, 1987.

———. *Tennōsei to Guntai* 天皇制と軍隊 [The Emperor System and Army]. Tokyo: Aoki, 1973.

———. *Uejini Shita Eirei-tachi* 飢死した英霊たち [The War Dead Who Died of Starvation]. Tokyo: Aoki Shoten, 2001.

Fussell, Paul. *Wartime: Understanding and Behavior in the Second World War II*. Oxford: Oxford University Press, 1989.

Gabriel, Richard A. *No More Heroes: Madness and Psychiatry In War*. New York: Hill and Wang, 1985.

Gluck, Carol. "The Idea of Showa." *Daedalus* 119, no. 3 (1990): 1–26. https://www.jstor.org/stable/20025314.

Greene, Bob. *Duty: A Father, His Son, and the Man Who Won the War*. New York: Perennial, 2001.

Grossman, Lt. Col. Dave. *On Combat: The Psychology and Physiology of Deadly Conflict in War and Peace*. New York: PPCT Research Publications, 2004.

———. *On Killing: The Psychological Cost of Learning to Kill in War and Society*. New York: Back Bay Books, 1995.

Griffin, Susan. *A Chorus of Stones: The Private Life of War*. New York: Anchor, 1992.

Handō Kazutoshi, et al. 半藤一利・保坂正康等. *Naze Ano Sensō ni Maketa no ka* なぜあの戦争に負けたのか [Why Did We Lose that War]. Tokyo: Bungei 2006.

Harada Keiichi 原田敬一. *Kokumingun no Shinwa: Heishi ni Narutoiukoto* 国民軍の神話 兵士になるということ [Myth of the Emperor's Soldier: What It Means to Be a Soldier]. Tokyo: Yoshikawa Hiroshi Bunkan, 2001.

Harries, Merion, and Susie Harries. *Soldiers of the Sun: The Rise and Fall of the Imperial Japanese Army.* New York: Random, 1991.

Harrison, Simon. "Skull Trophies of the Pacific War: Transgressive Objects of Remembrance." *Journal of the Royal Anthropological Institute* 12, no. 4 (2006): 817–36. http://www.jstor.org/stable/4092567.

Hartle, Anthony. "Atrocities in War: Dirty Hands and Noncombatants." *Social Research* 69, no. 4 (2002): 963–79. DOI: 10.1353/sor.2002.0052.

"Hayao Torao Gun'i Chūi Hōkokusho 'Senjō Shiinkeishō Sai ni Hanzai ni Tsuite'" 「早尾虎雄軍医中尉報告書「戦場神経症竝に犯罪に就て」」 [Army Doctor Lieutenant Torao Hayao's Report on Criminality Caused by War Fatigue]. http://d.hatena.ne.jp/gurugurian/19380401.

Hayashi Fumiko 林芙美子. *Hokugishi Butai* 北岸部隊 [Hokugishi Unit]. Tokyo: Chukou Bunko, 2002.

———. *Sensen* 戦線 [Battlefront]. Tokyo: Chukou Bunko, 2006.

Henmi Jun 辺見じゅん. *Senjō kara Todoita Ishō* 戦場から届いた遺書 [Last Wills That Arrived from the Battlefield]. Tokyo: Bunshun, 2003.

———. *Showa no Ishō: Minami no Senjō kara* 昭和の遺書 南の戦場から [Last Wills from Showa: From the Southern Battlefield]. Tokyo: Bunshun, 2002.

Henmi Masaaki 逸見勝克. *Gakudō Shūdan Sokaishi: Kodomo-tachi no Sentō Haichi* 学童集団疎開史 子供たちの戦闘配置 [Wartime Evacuation of School Children: The Strategic Positioning of Children]. Tokyo: Ootsuki Shoten, 1998.

Hikosaka Tai 彦坂諦. *Hito wa Dono yō ni Shite Hei to Naru no ka / Aru Munōheishi no Keiseki I, II* ひとはどのようにして兵となるのか―ある無能兵士の軌跡（上と下）[How Do Men Become Soldiers, Vol. 1 and 2). Tokyo: Keshi Shoubou, 1984.

———. *Gashi no Kenkyū: Gadarukanaru de Hei wa Ikanishite Shinda ka* 餓死の研究ガダルカナルで兵はいかにして死んだか [Study on Starvation: How Did Soldiers Die in Guadalcanal]. Tokyo: Rippu, 1992.

———. *Dansei Shinwa* 男性神話 [Masculine Mythology]. Tokyo: Komichi Shobō, 1991.

———. *Hei wa Dono yō ni shite Korosareru ka I, II* 兵はどのようにして殺されるか（上・下）(How Soldiers are Killed, Vol. I and II). Tokyo: Keshi Shobō, 1987.

Hiratsuka Masao, ed. 平塚柾緒編. *Shirarezaru Shōgensha-tachi: Heishi no Kokuhaku* 知られざる証言者たち 兵士の告白 [Unknown Witnesses: Confessions of Soldiers]. Tokyo: Shinjinbutsu Oorai Sha, 2007.

Holmes, Richard. *Acts of War: The Behavior of Men in Battle.* London: Cassells, 2003.

Honda Katsuichi 本田勝一. *Chūgoku no Tabi* 中国の旅 [Journey in China]. Tokyo: Asahi Bunko, 1981.

———. *Korosugawa no Ronri* 殺す側の論理 [Logics of the Killers]. Tokyo: Asahi Bunko, 1984.

———. *Tennō no Guntai* 天皇の軍隊 [The Army of the Emperor]. Tokyo: Asahi Bunko, 1991.

Hoshi Tōru 星徹. *Watashi-tachi ga Chūgoku de Shitakoto: Chūgokukikansha Renrakukai no Hitobito* 私たちが中国でしたこと：中国帰還者連絡会の人々 [What We Did in China: People from the China Returnee Group]. Tokyo: Ryokuryu, 2006.

Hosaka Masayasu 保坂正康. *Tokko to Nihonjin* 「特攻」と日本人 ["Special Attack" Corps and Japanese]. Tokyo: Kodansha, 2005.

"How Japs Fight." *Time Magazine* 41, no. 7 (1943): 26.

"How to Tell Your Friends from the Japs." *Time Magazine* 38, no. 25 (1941): 35.

Ibaraki Kyōdo Butai Shiryō Hozonkai 茨城郷土部隊史料保存会. *Ibaraki Rikugun Shobutai no Ato* 茨城陸軍諸部隊の跡 [Steps That the Ibaraki Battalions Took]. Ibaraki: Ibaraki Kyōdo Butai Shiryō Hozonkai, 1987.

Ibaraki-ken Shihen Sanshichōsanshibukai, ed. 茨城県史編さん市町村史部会編. *Ibarakikenshi — Shichōson, Hen III* 茨城県史：市町村編III [History of Ibaraki Prefecture: Cities, Towns and Villages, Vol. III]. Ibaraki: Ibaraki-ken, 1981.

Ibaraki-ken Tōbu Nyūginia Senyū-kai, ed. 茨城県東部ニューギニア戦友会. *Waga Tatakai no Ato: Daitoa Sensō Tōbu Nyūginia Sen* わが戦いの跡：大東亜戦争東部ニューギニア戦 [Traces of Our Battles: The Great Asia War, East New Guinea Battles]. Ibaraki: Ibaraki-ken Tobu Nyūginia Senyu-kai, n.d.

Ichinose Toshiya 一之瀬俊也. *Meiji, Taishō, Shōwa Guntai Manyuaru: Hito wa Naze Senjō e Itta ka* 明治・大正・昭和軍隊マニュアル　人はなぜ戦場へ行ったか [Military Manuals from Meiji, Taisho, and Showa Era: Why Did People Go to War]. Tokyo: Kobunsha, 2004.

———. *Jūgo no Shakaishi: Senshisha to Izoku* 銃後の社会史　戦死者と遺族 [Social History of Homefront: War Dead and His Family]. Tokyo: Yoshikawa Hiroshi, 2005.

———. *Kōgun Heishi no Nichijō Seikatsu* 皇軍兵士の日常生活 [Everyday Life of Soldiers of the Imperial Army]. Tokyo: Kodansha Gendai Shinsho, 2009.

Igarashi Yoshikuni 五十嵐恵邦. *Haisen no Kioku: Karada, Bunka, Monogatari, 1945–1970* 敗戦の記憶：身体・文化・物語 1945–1970 [Memory of the War Lost: Body, Culture, Narrative, 1945–1970]. Tokyo: Chuo Koron Shinsha, 2007.

Iida Susumu 飯田進. *Jigoku no Nihonhei: Nyūginia Sensen no Shinsō* 地獄の日本兵　ニューギニア戦線の真相 [Japanese Soldiers in Hell: Truth of the Battle in New Guinea]. Tokyo: Shinchō Shinsha 2008.

Iinuma Yoshie 飯沼佳恵. *Tai de Yuigon o Kake to Iwareta Keredomo* 隊で遺言を書けと言はれたけれども [The Unit Told Me to Write the Last Will But…]. Tokyo: Chuou Kōron Jigyō Shuppann, 2002.

Imamura, Shigeo. *The True Story of an American Kamikaze: A Memoir*. Tokyo: Sōshisha, 2003.

Inoue Toshikazu 井上寿一. *Nicchu Sensōka no Nihon* 日中戦争下の日本 [Japan During Japan-China War]. Tokyo: Kodansha Sensho Mechie, 2007.

Inoue Toshio 井上俊夫. *Hajimete Hito o Korosu: Rō-Nihonhei no Sensō-ron* 初めて人を殺す 老日本兵の戦争論 [I Killed a Man for the First Time: The Theory of War by an Aging Japanese Soldier]. Tokyo: Iwanami Shoten, 2005.

Ishikawa Tatsuzo 石川達三. *Ikiteiru Heitai* 生きている兵隊 [Soldiers Alive]. Tokyo: Chūkō Bunko, 2008.

Itō Keiichi 伊藤桂一. *Heitai-tachi no Rikugunshi* 兵隊たちの陸軍史 [Soldiers' History of the Military]. Tokyo: Shinchō Bunko, 2008.

———. *Wakaki Sedai ni Kataru Nichhū Sensō* 若き世代に語る日中戦争 [Japanese–Chinese War Stories to Tell the Younger Generation]. Tokyo: Bungei Shunju, 2007.

Iwata Shigenori 岩田重則. *Mura no Wakamono Kuni no Wakamono: Minzoku to Kokumin Sōgō* ムラの若者・くにの若者―民族と国民総合 [Young Men of the Village, Young Men of the Nation]. Tokyo: Mirai, 1996.

———. *Senshisha Reikon no Yukue Sensō to Minzoku* 戦死者霊魂のゆくえ　戦争と民族 [Where the Souls of the War Dead Go: War and Ethnicity]. Tokyo: Yoshikawa Hiroshi Bunkan, 2003.

Iwate-ken Nōson Bunka Kondan Kai, ed. 岩手県農村文化懇談会編集. *Senbotsu Nōmin Heishi no Tegami* 戦没農民兵士の手紙 [Letters of the Dead Farmer-Soldiers]. Tokyo: Iwanami Shinsho, 1961.

"Japanese Savagery." *The New York Times*, October 6, 1943: 22. https://www.nytimes.com/1943/10/06/archives/japanese-savagery.html.

Jeans, Roger B. "Alarm in Washington: A War Time 'Expose' of Japan's Biological Warfare Program." *The Journal of Military History* 71 (2007): 411–39. DOI: 10.1353/jmh.2007.0126.

Kamei Hiroshi 亀井宏. *Dokyumento Taiheiyō Sensō Zenshi Senjōtaikensha 300 nin Shuzai* ドキュメント 太平洋戦争全

史 戦場体験者300人取材 [The History of the Pacific War: Interview with 300 Soldiers]. Tokyo: Kodansha, 2009.

Kano Masanao 鹿野政直. *Heishi de Arukoto: Dōin to Jūgun no Seishinshi* 兵士であること 動員と従軍の精神史 [What It Means to Be a Soldier: Psychological History of Mobilization and Serving in Military]. Tokyo: Asahi Shimbunsha, 2005.

Katō Yōko 加藤陽子. *Sensō no Ronri Nichiro Sensō kara Taiheiyō Sensō made* 戦争の論理　日露戦争から太平洋戦争まで [Logic of War: From the Russo–Japanese War to The Pacific War]. Tokyo: Keisō Shobō 2005.

Katz, Fred Emil. *Ordinary People and Extraordinary Evil: A Report on the Beguilings of Evil.* New York: SUNY Press, 1993.

Kawamura Kunimitsu 川村邦光. *Seisen no Ikonogurafi Tennō to Heishi Senshisha no Zuzō Hyōshō* 聖戦のイコノグラフィ 天皇と兵士・戦死者の図像・表象 [Iconography of the Sacred War: Representation and Images of the Emperor, Soldiers, and the War Dead]. Tokyo: Seikyū-sha, 2007.

———. *Senshisha no Yukue Katari to Hyōshō kara* 戦死者のゆくえ 語りと表象から [Where the War Dead Go: Oral History and Representation]. Tokyo: Seikyū-sha, 2003.

———. *Minzoku Kūkan no Kindai: Wakamono, Sensō, Hisai, Takai no Fōkuroa* 民俗空間の近代　若者・戦争・被災・他界のフォークロア [Folkloric Space: Folklore of Young Men, Wars, Natural Disasters and the Otherworld]. Tokyo: Jōkyō Shuppan, 1996.

Kawano Hitoshi 河野仁. *Gyokusai no Guntai Seikan no Guntai: Nichibeiheishi ga Mita Taiheiyo Sensō* 玉砕の軍隊 生還の軍隊 日米兵士が見た太平洋戦争 [Army of No Surrender, Army of Staying Alive: The Pacific War as Seen by Japanese and American Soldiers]. Tokyo: Kodansha, 2001.

Kenmi Kazuo 計見一雄. *Sensō Suru Nō Hakyoku e no Byōri* 戦争する脳 破局への病理 [Brain at War: Disease toward Destruction]. Tokyo: Heibon-sha, 2007.

Kikuchi Keiichi 菊池敬一. *Nanasentsū no Gunjiyūbin* 7000 通の軍事郵便 [7000 Military Letters]. Tokyo: Hakujyu-sha, 1983.

———. *Nōmin Heishi no Koe ga Kikoeru* 農民兵士の声が聞こえる [I Can Hear the Voices of the Farmer-Soldiers]. Tokyo: Nihon Hōsō Shuppan Kyoukai, 1984.

Kitamura Riko 喜多村理子. *Chōhei: Sensō to Minshū* 徴兵・戦争と民衆 [Conscription: War and the Masses]. Tokyo: Yoshikawa Hiroshi Bunkan, 1999.

Kitamura Tsunenobu 北村恒信. *Senjiyōgo no Kiso Chishiki* 戦時用語の基礎知識 [Basic Dictionary of Wartime Vocabulary]. Tokyo: Kōjinsha NF Bunko, 2002.

Kobayashi Hideo 小林英雄. *Nicchu Sensō* 日中戦争 [Japanese–Chinese War]. Tokyo: Kodansha Gendai Shinsho, 2007.

Krus, David J., and Yoko Ishigaki. "Contributions to Psychohistory: XIX. Kamikaze Pilots: The Japanese versus the American Perspective." *Psychological Reports* 70 (1992): 599–602. DOI: 10.2466/pr0.1992.70.2.599.

Kudō Yukie 工藤雪枝. *Tokkō e no Rekuiemu* 特攻へのレクイエム [Requiem for the Special Forces]. Tokyo: Chuō Kōron Shinsha, 2001.

Kusaka Ruiko 草加類子. *Nyūginia: Kaerazaru Heishi "Senshi" no Giwaku o Otte* ニューギニア・帰らざる兵士「戦死」の疑惑を追って [New Guinea: A Soldier Who Didn't Return—Chasing after the Suspicion of "Death in Action"]. Tokyo: Shufu no Tomo, 1983.

Kuwahata Asami 桑畑朝海. *Dōkoku no Senjō* 慟哭の戦場 [Battlefield of Lament]. Tokyo: Bungeisha, 2000.

———. *Kaimetsu no Hohei Dainihyakusanjūnadan* 潰滅の歩兵第二百三十七連隊 [Infantry Unit 237 Destroyed]. Ibaraki: Yamano Shimbunsha Shuppan-bu (self-published), n.d.

Lebra, Takie Sugiyama. "The Social Mechanism of Guilt and Shame: The Japanese Case." *Anthropological Quarterly* 44, no. 4 (1971): 241–55. DOI: 10.2307/3316971.

Leshan, Lawrence. *The Psychology of War: Comprehending Its Mystique and Its Madness.* New York: Helios Press, 2002.

Majima Mitsuru 間嶋満. *Jigoku no Senjō Nyūginia Senki Sangaku Mitsurin ni Kieta Hiun no Gundan* 地獄の戦

場ニューギニア戦記 山岳密林に消えた悲運の軍団 [Hellish Battlefield: New Guinea — The Tragic Army Who Disappeared in Mountains and Jungles]. Tokyo: Kojinsha, 1985.

Matsuoka Tamaki, ed. 松岡環編. *Nankinsen Tozasareta Kioku o Tazunete Moto Heishi 120-nin no Shōgen* 南京戦 閉ざされた記憶を尋ねて 元兵士120人の証言 [Nanking Battle: Questioning the Closed Memory Confessions of 120 Former Soldiers]. Tokyo: Shakai Hyōron, 2002.

———. *Nankinsen Kirisakareta Junansha no Tamashii Higaisha 120-nin no Shōgen* 南京戦・切りさかれた受難者の魂: 被害者120人の証言 [Nanking Battle: Torn Souls of the Martyrs — 120 Eyewitness Accounts of the Victims]. Tokyo: Shakai Hyōron, 2004.

Matsutani Miyoko 松谷みよ子. *Guntai: Shōhei Kensa, Shinpei no Koro* 軍隊・徴兵検査・新兵のころ [Army: Conscription Examination, When We Were Still Grunts]. Tokyo: Chikuma Bunko, 2003.

Mikuni Ichirō 三國一朗. *Senchūyōgo-shū* 戦中用語集 [Dictionary of Words from the Wartime]. Tokyo: Iwanami, 1985.

Milgram, Stanley. *Obedience to Authority.* New York: HarperPerennial, 1974.

Military Intelligence Division, War Department. *The Punch Below the Belt: Japanese Ruses, Deception Tactics, and Antipersonnel Measures.* Special Series No. 33. Washington, DC: War Department, 1945. https://archive.org/details/ThePunchBelowTheBelt.

Mizuki Shigeru 水木しげる. *Mizuki Shigeru no Rabauru Senki* 水木しげるのラバウル戦記 [Mizuki Shigeru's Chronicle of War in Rabaul]. Tokyo: Chikuma, 1997.

Mizoguchi Akira 溝口章. *Senshi: Bōfu Guntai Techou Kou* 戦史 亡父軍隊手帳考 [Military History: My Dead Father's Army Journal]. Tokyo: Toyō Bijutsusha Shuppan Hanbai, 1995.

Moriyama Yohei, ed. 森山庸平編. *Beigun ga Kiroku Shita Nyūginia no Tatakai* 米軍が記録したニューギニアの戦

い [Battles in New Guinea as Recorded by the US Military]. Tokyo: Soushisha, 1995.

———. *Shōgen: Nankin Jiken to Sankō Sakusen* 証言 南京事件と三光作戦 [Eye Witness: Nanking Incident and Sankō Strategy]. Edited by Taiheiyou Sensou Kenkyukai. Tokyo: Kawade, 2007.

Morrison, Wayne. "'Reflections with Memories': Everyday Photography Capturing Genocide." *Theoretical Criminology* 8, no. 3 (2004): 341–58. DOI: 10.1177/1362480604044613.

Murase Moriyasu 村瀬守保. *Watashi no Jūgun Chūgoku Sensen: Ichiheishi ga Utsushita Senjō no Kiroku* 私の従軍中国戦線 一兵士が写した戦場の記録 [My Service in the China Front: Photographs of War front by a Soldier]. Tokyo: Nihon Kikanshi Shuppan, 2005.

Nadelson, Theodore. *Trained to Kill: Soldiers at War*. Baltimore: Johns Hopkins University Press, 2005.

Nanking Massacre Memorial, ed. 南京大虐殺遭難同胞記念館編 Minoru Kato, translation 加藤実訳. *"Kono Jijitsu o…" Nanking Daigyakusatsu Seizonsha Shōgenshū*「この事実を…」「南京大虐殺」生存者証言集 ["This Truth…": Collection of Testimonies of the Survivors of Nanking Great Massacre]. Nanking: Nanking University Press, 1994.

Nishimura Akira 西村明. *Shizume to Furui no Dainamizumu: Sengo Nihon to Sensō Shisha Irei* シズメとフルイのダイナミズム 戦後日本と戦争死者慰霊 [Dynamism of Exorcism and Drive: Post War Japan and Memorializing the War Dead]. Tokyo: Yushisha, 2006.

Nishimura Makoto 西村誠. *Taiheiyō Senseki Kikō Nyūginia* 太平洋戦跡紀行 ニューギニア [The Travelogue of Pacific War Former Battlefields: New Guinea]. Tokyo: Kōjinsha, 2006.

Nukuda Ichisuke 温田市助. *Chichi no Senki: Tōbu Nyūginia Sen* 父の戦記 東部ニューギニア戦 [My Father's War: Eastern New Guinea Battle]. Tokyo: Hatowa Suppan, 1987.

Oda Makoto 小田実. *Gyokusai* 玉砕 [Last Charges]. Tokyo: Shinchou, 1998.

Ōe Shinobu 大江志乃夫. *Chōheisei* 徴兵制 [Conscription System]. Tokyo: Iwanami Shinsho, 1981.

Ogawa Masatsugu 尾川正二. *Shi no Shima Nyūginia: Kyokugen no Nakano Ningen* 死の島　ニューギニア　極限の中の人間 [New Guinea Island of Death: Men in Extreme Situations]. Tokyo: Kōjinsha NF Bunko, 2004.

———. *Sensō: Hakō to Shinjitsu* 戦争　廃構と真実 [War: Myth and Truth]. Tokyo: Kōjinsha, 2000.

———. *Tōbu Nyūginia Sensen Suterareta Butai* 東部ニューギニア戦線　棄てられた部隊 [Battlefront in the Eastern New Guinea: The Abandoned Regiment]. Tokyo: Kōjinsha, 2000.

Okabe Makio, Ogino Fujio, and Yutaka Yoshida, eds. 岡部牧夫・荻野富士夫・吉田裕編. *Chūgokushinryaku no Shōgensha-tachi* 中国侵略の証言者たち [Perpetrators of Invasion of China]. Tokyo: Iwanami Shinsho, 2010.

Okumura Shoji 奥村正二. *Senjō Papua Nyūginia: Taiheiyōsensō no Sokumen* 戦場パプアニューギニア　太平洋戦争の側面 [Battlefield Papua New Guinea: Another Face of the Pacific War]. Tokyo: Chūkō Bunko, 1993.

Onda Shigetaka 御田重宝. *Tōbu Nyūginia Sen Shinkō Hen* 東部ニューギニア戦　進攻編 [Battles in the Eastern New Guinea: Advance]. Tokyo: Kodansha Bunko, 1988.

———. *Tōbu Nyūginia Sen Zenmetsu Hen* 東部ニューギニア戦　全滅編 [Battles in Eastern New Guinea: Complete Destruction]. Tokyo: Kodansha Bunko, 1988.

Ōnuki Emiko 大貫恵美子. *Gakutohei no Seishinshi Ataerareta Shi to Sei no Tankyū* 学徒兵の精神誌「与えられた死」と「生」の探求 [Psychological Study of Student-Soldiers: Examination of Forced Death and Life]. Tokyo: Iwanami, 2006.

——— [Ohnuki-Tierney, Emiko]. *Kamikaze Diaries: Reflections of Japanese Student Soldiers.* Chicago: University of Chicago Press, 2006.

Ozawa Makoto, and NHK Research Department 小澤真人＋NHK取材班. *Akagami: Otoko-tachi wa Kōshite Senjō e Okurareta* 赤紙　男たちはこうして戦場へ送られ

た [Akagami: How Men Were Sent to the War]. Tokyo: Sengensha, 1997.

Padover, Saul K. "Japanese Race Propaganda." *The Public Opinion Quarterly* 7, no. 2 (1943): 191–204. DOI: 10.1086/265613.

Penney, Matthew. "Far from Oblivion: The Nanking Massacre in Japanese Historical Writing for Children and Young Adults." *Holocaust and Genocide Studies* 22, no. 1 (2008): 15–48. DOI: 10.1093/hgs/dcn003.

Prevost, Ann Marie. "Race and War Crimes: The 1945 War Crimes Trial of General Tomoyuki Yamashita." *Human Rights Quarterly* 14, no. 3 (1992): 303–38. DOI: 10.2307/762369.

Price, Willard. "The Men Who Drive Japan." *Harper's Magazine* 186, no. 1111 (1942): 47–57.

Rees, Laurence. *Their Darkest Hour: People Tested to the Extreme in WWII*. London: Ebury Publishing, 2007.

Rice, Geoffrey W., and Edwina Palmer. "Pandemic Influenza in Japan, 1918–19: Mortality Patterns and Official Responses." *Journal of Japanese Studies* 19, no. 2 (1993): 389–420. DOI: 10.2307/132645.

Roth, Paul A. "Hearts of Darkness: 'Perpetrator History' and Why There Is No Why." *History of the Humane Sciences* 17, nos. 2–3 (2005): 211–51. DOI: 10.1177/0952695104047303.

Saeki Hiromi 佐伯廣美. *Nyūginia Chinkonki* ニューギニア鎮魂記 [New Guinea Requiem]. Tokyo: Kindai Bungei Sha, 1994.

Sakuramoto Tomio 桜本富雄. *Gyokusai to Kokusō: 1943-nen 5-gatsu no Shisō* 玉砕と国葬—1943年5月の思想 [Gyokusai and National Funeral: Ideologies of May 1943]. Tokyo: Kaisai-sha, 1984.

"Sample of Japanese Imperial Mysticism." *Harper's Magazine* 186, no. 1112 (January 1943): 191.

Sano, Iwao Peter. *One Thousand Days in Siberia: The Odyssey of a Japanese–American POW*. Lincoln: University of Nebraska Press, 1998.

Sasaki Motokatsu 佐々木元勝. *Yasen Yūbin Hata Nicchū Sensō ni Jūgun Shita Yūbin Kyokuchō no Kiroku* 野戦郵便旗 日中戦争に従軍した郵便長の記録 [The Field Postal Flag: A Postal Master's Record During the Japan–China War]. Tokyo: Gendaishi Shuppan-kai, 1973.

Satō Kihohiko 佐藤清彦. *Dotanba ni Okeru Ningen Kenkyū: Nyūginia Yami no Senato* 土壇場における人間研究 ニューギニア 闇の戦跡 [A Study on Men on Edge: New Guinea, The Dark Battlefield]. Tokyo: Fuyōshōbō Shuppan, 2003.

Satō Shōdō 佐藤正導. *Nicchū Sensō: Aru Wakaki Jūgunsō no Shuki* 日中戦争・ある若き従軍僧の手記 [Sino-Japanese War: A Chronicle of a Military Monk]. Tokyo: Nihon Arumito-sha, 1992.

Satō Tadao 佐藤忠男. *Kusa no Ne no Gunkoku Shugi* 草の根の軍国主義 [Grassroot Militarism]. Tokyo: Heibon, 2007.

Seaton, Philip. "'Do You Really Want to Know What Your Uncle Did?' Coming in Terms with Relatives' War Actions in Japan." *Oral History* 34, no. 1 (2006): 53–60. https://www.jstor.org/stable/40179844.

Senda Kakō 千田夏光. *Kinjirareta Senki: Nyūginia, Gatō, Ruson, Kigachitai* 禁じられた戦記：ニューギニア、ガ島、ルソン・飢餓地帯 [The Forbidden War Memory: New Guinea, Gaudalcanal, Luson, the Hunger Islands]. Tokyo: Sekibun, 1975.

———. *"Seisen" no Na no Moto ni* 「聖戦」の名のもとに [In the Name of the Holy War]. Tokyo: Rōdoushuhō, 1995.

Sereny, Gitta. *Into That Darkness: An Examination of Conscience.* New York: Vintage, 1983

Sheftall, M.G. *Blossoms in the Wind: Human Legacies of the Kamikaze.* New York: Nal Caliber, 2006.

Sheldon, Charles. D. "Japanese Aggression and the Emperor, 1931–1941, from Contemporary Diaries." *Modern Asian Studies* 10, no. 1 (1976): 1–40. DOI: 10.1017/S0026749X00013342.

Shigematsu Kiyoshi, and Kō Watanabe 重松清・渡辺考. *Saigo no Kotoba: Senjō ni Nokosareta Nijūyonmanji no Todokanakatta Tegami* 最後の言葉 戦場に遺された二十

四万字の届かなかった手紙 [Last Words: 240,000 Letters Left Behind in the Battlefield That Never Reached Home]. Tokyo: Kodansha, 2004.

Shimada Kakuo 島田覚夫. *Watashi wa Makyō ni Ikita: Shūsen mo Shirazu Nyūginia no Yamaoku de Genshiseikatsu 10nen* 私は魔境に生きた 終戦も知らずニューギニアの山奥で原始生活10年 [I Lived in Hell: 10 Years of Living Primitively in New Guinea Without Knowing That the War Had Ended]. Tokyo: Kōjinsha NF Bunko, 2007.

Shimamoto Yasuko 島本慈子. *Sensō de Shinuto Iukoto* 戦争で死ぬ、ということ [What It Means to Die in War]. Tokyo: Iwanami, 2006.

Shimizu Hiroshi, ed. 清水寛編. *Nihonteikoku Rikugun to Seishinshōgaiheishi* 日本帝国陸軍と精神障害兵士 [Japanese Imperial Army and Psychiatric Soldiers]. Tokyo: Fuji, 2007.

Shimizu Mitsuo 清水光雄. *Seigo no Kōgun Heishi: Kuhaku no Toki, Senshōbyōtō kara* 最後の皇軍兵士 空白の時、戦傷病棟から [The Last Imperial Soldiers: The Passing Time, From the Veterans' Hospital]. Tokyo: Gendai Hyōron-sha, 1985.

Shimokawa Kōshi 下川耿史. *Nihon Zankoku Shashinshi* 日本残酷写真史 [A History of Japanese Brutal Photography]. Tokyo: Sakuhin-sha, 2006.

———. *Shitai no Bunkashi* 死体の文化史 [A Cultural History of Dead Bodies]. Tokyo: Seiyumi-sha, 1990.

Shōguchi Yasuhiro 将口泰浩. *Kaerazaru Hito Mikikanhei 62-nenme no Shōgen* 帰らざる人 未帰還兵　62年目の証言 [Unreturned Men: Unrepatriated Soldiers — Confessions of the 62nd Year]. Tokyo: Sankei Shimbunsha no Hon, 2008.

"Shōwa 16 Nen 12 Gatsu 8 Nichi no Radio: Gozen 7 Ji no Rinji Nyūsu Taiheiyō Sensō Kaisei"「昭和16年12月8日のラジオ（一）　午前7時の臨時ニュース「太平洋戦争開戦」[December 8, 1941 Radio Broadcast: The Start of the Pacific War]. *NHK Archives*. https://www2.nhk.or.jp/archives/movies/?id=D0001400316_00000. Accessed June 20, 2025.

Shūkan Asahi, ed. 週刊朝日編. *Chichi no Senki* 父の戦記 [My Father's War]. Tokyo: Asahi Bunko, 1965, 2008.

Sledge, E.B. *With the Old Breed: At Peleliu and Okinawa.* New York: Ballantine, 2007.

Sledge, Michael. *Soldier Dead: How We Recover, Identity, Bury and Honor Our Military Fallen.* New York: Columbia University Press, 2005.

Sontag, Susan. *Regarding the Pain of Others.* New York: Picador, 2003.

Staub, Ervin. *The Roots of Evil: The Origins of Genocide and Other Group Violence.* Cambridge: Cambridge University Press, 1989.

Stern, Paul C. "Why do People Sacrifice for Their Nations?" *Political Psychology* 16, no. 2 (1995): 217–35. DOI: 10.2307/3791830.

Sugano Shigeru 菅野茂. *7% no Unmei: Tōbu Nyūginia Sensen Mitsurin kara no Seikan* 7%の運命 東部ニューギニア戦線 密林からの生還 [7% Survival Rate: The Eastern New Guinea Battle Front, Safe Return from the Jungle]. Tokyo: Kōjinsha NF Bunko, 2006.

Surin, Kenneth. "Conceptualizing Trauma, but What About Asia?" *Positions: East Asia Cultures Critique* 16, no. 1 (2008): 15–37. DOI: 10.1215/10679847-2007-010.

Suzuki Masami 鈴木正巳. *Nyūginia Gun'e Senki Jikoku no Senjō o Ikinuita Ichiguni no Kiroku* ニューギニア軍医戦記 地獄の戦場を生きた一軍医の記録 [The Medical Officer's War Tales in New Guinea]. Tokyo: Kōjinsha NF Bunko, 2001.

Suzuki Shōshichi 鈴木正七. *Fukuinsen ga Yattekita Waga Chūgoku Nyūginia Senki* 復員船がやってきた わが中国・ニューギニア戦記 [The Repatriation Boat: My War in China and New Guinea]. Tokyo: Bungeisha, 2005.

Tajima Kazuo 田島一夫. *Taiken Jitsuroku to Kenshō: Nyūginia Sen Higeki no Kyōmei to Kenshō* 体験実録と検証 ニューギニア戦 悲劇の究明と検証 [New Guinea Battle: Investigation and Inspection of Tragedy]. Tokyo: Senshi Kankou Kai, 1994.

Takada Eriko 高田里恵子. *Gakureki Kaikyū Guntai: Kōgakurekiheishi-tachi no Yuutsu na Nichijō* 学歴・階級・軍

隊 高学歴兵士たちの憂鬱な日常 [Education, Class, and Military: Depressing Days of the Educated Soldiers]. Tokyo: Chūkō Shinsho, 2008.

Takahashi Hikohiro 高橋彦博. *Minshū no Gawa no Sensō Sekinin* 民衆の側の戦争責任 [War Responsibility of the Masses]. Tokyo: Aoki, 1989.

Talbott, John E. "Soldiers, Psychiatrists, and Combat Trauma." *Journal of Interdisciplinary History* 27, no. 3 (1997): 437–54. DOI: 10.2307/205914.

Tanaka Nobumasa 田中伸尚. *Dokyumento Yasukuni Soshō: Senshisha no Kioku wa Dare no Mono ka* ドキュメント 靖国訴訟 戦死者の記憶はだれのものか [Document: Yasukuni Court Cases: Whom Does the Memory of the War Dead Belong To?]. Tokyo: Iwanami, 2007.

Tanaka Toshiyuki 田中利幸. *Sensō Hanzai no Kōzō: Nihongun wa Naze Minkanjin o Koroshitanoka* 戦争犯罪の構造：日本軍はなぜ民間人を殺したのか [Construction of War Crimes: Why Did the Japanese Army Kill Civilians]. Tokyo: Ootsuki, 2007.

Tanakamaru Katsuhiko 田中丸勝彦. *Samayoeru Eirei-tachi: Kuni no Mitama, Ie no Hotoke* sさまよえる英霊たち 国のみたま 家のほとけ [Wandering War Dead: Gods of the Nation, Ancestors of Home]. Tokyo: Kashiwa, 2002.

Toyotani Hidemitsu 豊谷秀光. *Nyūginia Donpeiroku Jigoku Angyo* ニューギニア鈍兵録　地獄行脚 [A Story of the Worst Soldier in New Guinea: Pilgrimage through Hell]. Tokyo: Nihon Tosho Hakkou Kai, 1996.

Tsuboi Hideto 坪井秀人. *Sensō no Kioku o Sakanoboru* 戦争の記憶をさかのぼる [Tracing Back the Memory of War]. Tokyo: Chikuma Shinsho, 2005.

Tsuda Michio 津田道夫. *Nankin Daigyakusatsu to Nihonjin no Seishin Kōzō* 南京大虐殺と日本人の精神構造 [Nanking Massacre and the Psychological Make-up of Japanese]. Tokyo: Shakai Hyōron-sha, 1995.

Tsuge Hisayoshi 柘植久慶. *Senjō o Kakenuketa Shokan* 戦場を駆け抜けた書簡 [Military Posts through the Ages]. Tokyo: Hara Shobō, 1991.

Uematsu Jinsaku 植松仁作. *Seibu Nyūginia: Senki Chi no Hate ni Shisu* 西部ニューギニア戦記 地の果てに死す[The War on Western New Guinea: They Died At the End of the World]. Tokyo: Tōsho Shuppan, 1976.

VAWW-NET Japan, ed. バウネット・ジャパン編. *Kagai no Seishin Kōzō to Sengo Sekinin* 加害の精神構造と戦後責任 [The Psychological Construction of Perpetration and Post-War Responsibility]. Tokyo: Rokyufu, 2000.

Wakabayashi, Bob Tadashi. "The Nanking 100-Man Killing Contest Debate: War Guilt Amid Fabricated Illusions, 1971–75." *Journal of Japanese Studies* 26, no. 2 (2000): 307–40. DOI: 10.2307/133271.

Warren Kozak, *The Rabbi of 84th Street: The Extraordinary Life of Haskel Besser.* New York: HarperCollins, 2004.

Waller, James. *Becoming Evil: How Ordinary People Commit Genocide and Mass Killing.* Second edition. Oxford: Oxford University Press, 2007.

Weingartner, James J. "Trophies of War: U.S. Troops and the Mutilation of Japanese War Dead, 1941–1945." *The Pacific Historical Review* 61, no. 1 (1992): 53–67. DOI: 10.2307/3640788.

Yamamoto Shichihei 山本七平. *Watashi no Naka no Nihongun* 私の中の日本軍 [Japanese Military Inside of Me]. Tokyo: Bunshun Bunko, 2005.

Yomiuri Newspaper Osaka Society Department, ed. 読売新聞大阪社会部編. *Nyūginia: Shibunkisha ga Kataritsugu Sensō, 4* ニューギニア　新聞記者が語り継ぐ戦争4[New Guinea: War as Told by Journalists, Vol. 4]. Osaka: Shinpu Shobō, 1988.

Yon'ichi-kai Honbu 41会本部. *Dai 41 Shidan Nyūginia Sakusen Shi* [41th Army History of New Guinea Battle]. Self Published.

Yon'ichi-kai Jimukyoku 四一会事務局. *Yon'ichi-kai Kaishi Daichigō* 四一会会誌第一号から第8号 [41st Regiment Journal]. Numbers 1–8. Tokyo: Yon'ichi-kai, 1956, 1957, 1957, 1958, 1959, 1960, 1961, and 1962.

———. *Yon'ichi-kai Kaishi 35-shūnenn Kinengō* 四一会会誌35周年記念号 [41st Regiment Journal 35th Anniversary Edition]. Tokyo: Yon'ichi-kai, 1989.

———. *Yon'ichi-kai Kaishi 40-shūnen Kinengō* 四一会会誌40周年記念号 [41st Regiment Journal 40th Anniversary Edition]. Tokyo: Yon'ichi-kai, 1994.

www.ingramcontent.com/pod-product-compliance
Lightning Source LLC
Chambersburg PA
CBHW050103170426
43198CB00014B/2444